A Laboratory Manual on Plant Pathology

Biochemical and Molecular Techniques

The Authors

Sumit Kumar did his graduation B.Sc. (Ag.) Hons. from Sardar Vallabhbhai Patel University of Agri. and Tech. Meerut. He has completed M.Sc. (Ag.) in Plant pathology from Chandrashekhar Azad University of Agri. and Tech., Kanpur. He has qualified ICAR-NET with 65 per cent. Presently, he is pursuing Ph.D. from Department of Mycology and Plant Pathology, Institute of Agricultural Sciences, Banaras Hindu University, Varanasi. Now he is an awardee of UGC Research Fellowship. He has published many research papers in renowned journal and many book chapters in reputed books. Furthermore, he has also published 11 popular articles in different magazine that helpful for growers and science readers.

Lopamudra Behera did her graduation B.Sc. (Ag) Hons. and M.Sc. (Ag) in Plant Pathology from Odisha University of Agriculture and Technology, Bhubaneswar, Odisha. Presently, she is pursuing Ph.D. from Department of Mycology and Plant Pathology, Institute of Agricultural Sciences, Banaras Hindu University, Varanasi. She has qualified ICAR-NET. She has been awarded with UGC Research Fellowship. She has published many research papers in renowned journal and many book chapters and more than 10 popular articles in different magazines.

Anmol Singh Yadav did his graduation B.Sc. (Ag.) from the Acharya Narendra Deva University of Agri. and Tech., Ayodhya in the year 2014. He obtained his M.Sc. (Ag.) in Plant pathology from the Chandra Shekhar Azad University of Agri. and Tech. Kanpur. He has qualified ICAR-NET. He is an awardee of the UGC- Research fellowship, at present; he is a Ph. D. research scholar in the Department of Mycology and Plant Pathology, Institute of Agricultural Sciences, Banaras Hindu University, Varanasi. His field of research is host-pathogen interaction and plant disease resistance. He has published many scientific popular articles in various magazines, book chapters in reputed books.

Dr. Vinod Kumar Sonakar did his graduation B.Sc. (Ag.) Hons. from Udai Pratap Autonomous College, Varanasi affiliated to V.B.S.P.U. Jaunpur. He has completed M.Sc. (Ag.) in Plant Pathology from Chandra Shekhar Azad University of Agri. and Tech., Kanpur and Ph.D. in Plant Pathology from Banaras Hindu University, Varanasi. He has qualified ICAR-NET in 2013. He was awarded Rajiv Gandhi National Fellowship and DST Inspire Fellowship. Presently, he has working as Technical Assistant in Agriculture Department, Govt. of Uttar Pradesh. He has published many research papers and articles in renowned journal and magazine.

A Laboratory Manual on Plant Pathology
Biochemical and Molecular Techniques

– *Authors* –

Sumit Kumar

Lopamudra Behera

Anmol Singh Yadav

Dr. Vinod Kumar Sonakar

BIOTECH

2022

BIOTECH BOOKS®

ISBN 978-81-7622-511-3

Published by: **BIOTECH BOOKS®**
4762-63/23, Ansari Road, Darya Ganj,
New Delhi - 110 002
Phone: 09899295259, 011-43003222
E-mail: biotechbooks@yahoo.co.in
Website: www.biotechbooks.in

Typeset at: **Classic Computer Services**
Delhi - 110 035

Printed at: **Replika Press Pvt. Ltd.**

PRINTED IN INDIA

Southern Federal University, Russian Federation

Dr. Vishnu D. Rajput
M.Sc. Ag, PhD, PDF,
Leading Researcher,
(Highly Qualified Specialist)
Southern Federal University,
Rostov-on-Don, Russia
E-mail: rvishnu@sfedu.ru
rajput.vishnu@gmail.com
Mob: +7-9185890093
https://www.researchgate.net/profile/Vishnu_Rajput2

FOREWORD

Scientific knowledge on plant pathology is rapidly increasing; however, the practical information is often sacred. This manual is a compilation of all aspects of updated biochemical and molecular techniques of plant pathology.

The techniques described in the manual are self-explanatory so the students including other disciplines could also benefit. I appreciate the effort of the authors for the attempt made for the upliftment of the students, researchers, and teachers' community.

I appreciate authors for bringing out this most needed and useful practical manual entitled "**A Laboratory manual on Plant Pathology** (Biochemical and molecular techniques)" to students, researchers and teachers. Congratulations to **S. Kumar**, **L. Behera**, **A. S. Yadav** and **V. K. Sonakar**, for this useful updated manual, for the accomplishment of this job and hope that they will continue to significantly contribute to the body of scientific knowledge, create a better awareness of the challenging problems and develop trained and qualified manpower in the field of Plant Pathology.

Yours faithfully

(Vishnu D. Rajput)

Date: 24.06.2021
Place: Rostov, Russia

Acknowledgements

It is by the immense blessing of the Almighty for his grace and blessings showered on us in each and every step of our life, that we could achieve this goal.

We are inexpressibly ecstatic to extend our deep sense of gratitude and sincere thanks to our advisor Dr. Ram Chandra, Professor, Department of Mycology and Plant Pathology, Institute of Agricultural Sciences, Banaras Hindu University for his sagacious guidance, constant inspiration, encouragement and full support in planning and execution of this work. He had been very kind to us in extending all possible helps and co-operation in writing and compilation of the manual. His patience and persistence became an ideal for us.

We sincerely accentuate our everlasting gratitude to the esteemed teachers of Department of Mycology and Plant Pathology, Institute of Agricultural Sciences, Banaras Hindu University for their valuable suggestions, constant support, constructive criticism, unconditional help and encouragement during preparation of this practical manual.

The words can hardly express our feelings for inspiration, incessant encouragement, blessings and unabated assistance of all kind which we derived from our parents and family members for their unconditional love and moral support.

From the special corner of my heart, I would like to express my thanks to my best friend Mani Tiwari for the helping and supporting.

With utmost satisfaction we acknowledge the enormous help of our colleagues and our beloved junior Manjari Mishra and Vanama for their valuable inputs and suggestions in making the practical manual diverse, comprehensive and according to the needs of the hour.

I would like to present a bouquet of thanks to my father Mr. Kailash Kumar Gupta, my mother Mrs. Gomati Gupta, my sisters Pragati Gupta, for their unconditional love and never-ending care and support.

Further suggestions towards improvement of this manual are welcome.

Sumit Kumar

Lopamudra Behera

Anmol Singh Yadav

Dr. Vinod Kumar Sonakar

Preface

Amendments and evolution in science and technology come about through science. The blueprints of this lab manual is simple yet deals with experiments that are translational with different chapters dealing with methods and experiments of basic biochemistry and molecular biology. The principle behind each experiment is defined so that a student can understand the concepts underlining each experiment. Since we have made individual experiment independent, students and researchers can start from any experiment. Each experiment starts with an introduction explaining the experiments followed by basic principle behind it. This is followed by materials and chemicals required to conduct the experiment and step-by-step procedure. The experiment ends with the precautions that need to be taken care of during the experiment to conduct the experiment safely. The chapter also includes calculation and some pictures towards the end of the experiment. This manual is divided into two parts. The first part which includes chapter 1 and 2 discusses different methods of biochemical analysis. The second part focuses on different molecular techniques.

This laboratory manual has been designed for post graduate students, researchers and others professionals of Plant Pathology and allied disciplines. This practical manual makes a very valuable contribution to agriculture and crop sciences. This work would not have been possible without the invaluable contribution, knowledge and expertise of all the authors.

Sumit Kumar

Lopamudra Behera

Anmol Singh Yadav

Dr. Vinod Kumar Sonakar

Contents

Chapter 1

Methods of Biochemical Analysis

1. PEROXIDASE (PO) ACTIVITY

Introduction

Peroxidases are the ubiquitous enzyme and reported to be present in all living genera. PO can catalyze the oxidation of various organic and inorganic substrates by reacting with hydrogen peroxide and similar molecules. Due to its wide catalytic activity, peroxidase can act in the removal of both phenolic compounds and peroxides, in chemical synthesis and, according to recent studies, in mycotoxin degradation.

Materials Required

- ☆ Fresh inoculated and controlled leaf sample
- ☆ Mortar and pestle
- ☆ Centrifuge
- ☆ UV-visible spectrophotometer

Chemicals Required

- ☆ 100 mM potassium phosphate buffer (pH 7.5)
- ☆ 0.1 M phosphate buffer (pH 7.0)
- ☆ 20 mM guaiacol solution
- ☆ 12.3 mM hydrogen peroxide (H_2O_2)

Procedure (by Sadasivam and Manickam, 1996)

1. Enzyme extract for peroxidase is prepared by grinding 0.5 g of controlled and inoculated leaf samples separately with 2 mL of 100 mM potassium phosphate buffer (pH 7.5). The homogenate is centrifuged at 15,000 rpm for 15 min at 4°C and the supernatant is used as enzyme source.

2. For the peroxidase assay, the reaction mixture is prepared by adding 2.85 mL 0.1 M phosphate buffer (pH 7.0), 50 µl of 20 mM guaiacol solution, 50 µl of 12.3 mM hydrogen peroxide and 50 µl of enzyme extract. The reaction is allowed to proceed for 3 min.
3. Absorbance at 470 nm is measured 30 seconds after adding the enzyme extract to the substrate, and change in the absorbance is recorded up to 3 min.
4. The peroxidase activity is determined using molar absorption coefficient U = 26.6 mM^{-1} cm^{-1}.

Precautions

1. All buffers/solutions should be prepared using Millipore water and stored at 4°C until they are used and they are brought to room temperature before the experiment.
2. The crude leaf extract should be prepared fresh daily from leaf tissue, kept on ice and all assays must be conducted within a few hours.
3. For all assays, care must be taken to avoid air bubbles forming in the cuvettes before making absorbance measurements.

2. POLYPHENOL OXIDASE (PPO)

Introduction

Polyphenol oxidases (PPO) are copper-containing monooxygenases catalyzing the o- hydroxylation of phenols (cresolase activity) and the oxidation of o-diphenols (catecholase activity) to the corresponding o-quinones. These enzymes are often called tyrosinases, phenolases, phenol oxidases or polyphenol oxidases. These enzymes have been found in fungi, higher plants, arthropods, amphibians, mammals, and show different substrate specificities.

Assay is based on the formation of quinones from phenolic compounds, which exhibited non enzymatic reactions to develop colour end compounds. The methods actually used to determine the activity of PPOs cannot permit to separate and distinguish between enzymatic and non-enzymatic steps of phenolic compounds oxidation.

Materials Required

- ✰ Fresh controlled and inoculated leaf samples
- ✰ Mortar and Pestle
- ✰ Centrifuge
- ✰ UV- visible spectrophotometer
- ✰ 0.1 M sodium phosphate buffer (pH 6.5)
- ✰ 0.01 M catechol

Procedure

1. Enzyme extract for polyphenol oxidase is prepared by grinding 1 g of controlled and inoculated leaf samples separately with 2 mL of 0.1 M sodium phosphate buffer (pH 6.5). The homogenate is centrifuged at 16,000 rpm for 15 min at 4°C and the supernatant is used as enzyme source.
2. For the PPO assay, the reaction mixture is prepared by adding 1.5 mL of 0.1 M sodium phosphate buffer (pH 6.5), 200 µl of 0.01 M catechol and 200 µl enzyme extract. The reaction is allowed to proceed for 3 min.
3. Absorbance at 495 nm is measured 30 seconds after adding the enzyme extract to the substrate, and change in the absorbance is recorded up to 3 min.
4. The enzyme activity is expressed as changes in absorbance at 495 nm min^{-1} g^{-1} protein.

Calculation

Activity U/mL = Abs at 3 min - Abs at 0 min X total reaction volume/Time interval X 0.2

Reaction vol 1.9 Time interval = 3min

0.2 mL is volume of enzyme taken in reaction mixture

Precautions

1. The crude leaf extract should be prepared fresh daily from leaf tissue.
2. Care must be taken to avoid air bubbles forming in the cuvettes before making absorbance measurements.

3. PROTEIN

A. Estimation of Protein by Lowry's

Principle

The principle behind the Lowry method of determining protein concentrations lies in there activity of the peptide nitrogen[s] with the copper [II] ions under alkaline conditions and the subsequent reduction of the Folin- Ciocalteay phosphomolybdic phosphotungstic acid to heteropoly molybdenum blue by the copper-catalyzed oxidation of aromatic acids. The Lowry method is sensitive to pH changes and therefore the pH of assay solution should be maintained at 10 - 10.5.

The Lowry method is sensitive to low concentrations of protein. The major disadvantage of the Lowry method is the narrow pH range within which it is accurate. A variety of compounds will interfere with the Lowry procedure. These include some amino acid derivatives, certain buffers, drugs, lipids, sugars, salts, nucleic acids and sulphydryl reagents (Dunn, 1992). Price (1996) notes that ammonium ions, zwitter ionic buffers, nonionic buffers and thiol compounds may also interfere with the Lowry reaction. These substances should be removed or diluted before running Lowry assays.

Chemicals Required

A. 2 per cent Na_2CO_3 in 0.1 N NaOH
B. 1 per cent NaK Tartrate in H_2O
C. 0.5 per cent $CuSO_4$. 5 H_2O in H_2O
D. Reagent I- 48 mL of A + 1 mL of B + 1 mL C
E. Reagent II- 1-part Folin-Phenol [2 N]: 1 part water

Procedure

1. For estimation of protein, first of all 0.2 mL of BSA working standard is taken in 5 test tubes and volume is made up to 1mL by adding distilled water and the test tube with 1 mL distilled water is served as blank.
2. Then 4.5 mL of reagent-I is added and incubated for 10 minutes.
3. After incubation 0.5 mL of reagent II is added and incubated for 30 minutes.
4. Absorbance is measured at 660 nm and standard graph is plotted.
5. The amount of protein present in the sample is estimated from the standard graph.

B. Estimation of Proteins by Bradford

Principle

The protein in solution can be measured quantitatively by different methods.

The methods described by Bradford uses a different concept- the protein's capacity to bind to a dye, quantitatively. The assay is based on the ability of proteins to bind to Coomassie brilliant blue and form a complex whose extinction coefficient is much greater than that of free dye.

Materials Required

- ☆ Test tubes
- ☆ Graduated cylinder
- ☆ Weight Balance
- ☆ UV spectrophotometer

Chemicals Required

- ☆ 100 mg of Coomassie-Brilliant blue G250 is dissolved in 50 mL of 95 per cent Ethanol.
- ☆ 100 mL of 85 per cent phosphoric acid is added and volume is made up to 600 mL by adding distilled water.
- ☆ After filtering the solution, 100 mL of glycerol is added and then volume is made up to 1000 mL.
- ☆ The solution can be used after 24 hrs.
- ☆ BSA

Procedure

1. Various concentrations of standard protein solutions are prepared from the stock solution (say 0.2, 0.4, 0.6, 0.8 and 1.0 mL) into series of test tubes and the volume is made up to 1 mL in each test tube.
2. Then 0.2 mL of the sample is pipette out in two other test tubes and the volume is made up to 1 mL.
3. A tube with 1 mL of water is served as blank.
4. About 5.0 mL of Coomassie brilliant blue is added to each tube and mixed by vortex or inversion.
5. Test tubes are kept for 10-30 minutes and absorbance of each sample is measured at 595 nm wavelength.
6. Graph is plotted by using the absorbance of the standards verses their concentrations.
7. Another graph of optical density versus concentration is plotted. From graph amount of protein in unknown sample is estimated.

4. TOTAL PHENOLIC CONTENT (TPC)

Introduction

The content of total phenolic compounds of different plants is determined by Folin- Ciocalteau Reagent (FCR). The FCR actually measures a sample's reducing capacity. The exact chemical nature of the FC reagent is not known, but it is believed to contain hetero polyphosphotungstates- molybdates. Sequences of reversible one or two electron reduction reactions lead to blue product, possibly $(PMoW_{11}O_{40})_4$. In essence, it is believed that the molybdenum is easier to be reduced in the complex and electron transfer reaction occurs between reductants and Mo (VI):

$$\text{Mo (VI)} + e^- \rightarrow \text{Mo (V)}$$

Materials Required

- ✰ Trichloro acetic acid (0.1 per cent)
- ✰ Folin- Ciocalteu reagent (10 per cent)
- ✰ Sodium carbonate (7.5 per cent)
- ✰ Gallic acid
- ✰ Centrifuge
- ✰ Spectrophotometer

Procedure

1. For estimation of TPC, leaf tissue extract is prepared by grinding 1.0 g of fresh tissues into fine powder.
2. To this, 20 mL of 0.1 per cent TCA (Trichloro acetic acid) is added and the content is centrifuged at 12,000 rpm for 10 min at room temperature (RT). The supernatant thus obtained is collected for estimation of TPC.
3. For this, 0.5 mL of the aqueous extract is added to 2.5 mL of 10 per cent Folin- Ciocalteu reagent (v/v) and 2 mL of 7.5 per cent sodium carbonate.
4. The reaction mixture is incubated at 45°C for 40 min and the absorbance was measured at 765 nm using spectrophotometer.
5. Gallic acid is used as standard phenol. The TPC is expressed as milligrams of gallic acid equivalents per gram tissue extract.

Calculation

$$C = c\,V/m$$

where,

C = Total phenolic content mg GAE/g dry extract

c = Concentration of gallic acid obtained from calibration curve in mg/mL
V = volume of extract in mL

m = Mass of extract in gram

5. SUPEROXIDE DISMUTASE (SOD)

Introduction

Adverse environmental factors result in increased levels of reactive oxygen species (ROS) that are detrimental to the plant. To avoid damage caused by these excess ROS, plants have developed elaborate mechanisms to manage them at sustainable levels. SODs are considered the first line of defense against damage by the superoxide radical. These are metalloproteins that catalyze the dismutation reaction of two superoxide free radicals to one molecule of O_2 and one molecule of H_2O_2. SODs occur in different isoforms with different metal cofactors.

Assay is based on reduction of Nitroblue tetrazolium (NBT), which is a yellow compound to blue monoformazon by the superoxide radical. SOD activity is quantified based on the competitive inhibition of NBT reduction by the superoxide radical. At near- neutral pH, the dismutation rate of superoxide radicals increases 1,000-fold in the presence of SOD.

Materials Required

- ☆ Fresh leaf sample
- ☆ Mortar and pestle
- ☆ Centrifuge
- ☆ 15 W fluorescent lamps: 2
- ☆ UV-visible spectrophotometer

Chemicals Required

- ☆ 100 mM Potassium phosphate buffer (pH 7.5)
- ☆ 200 mM Methionine
- ☆ 0.5 mM and 3 mM Ethylene-diamine tetra acetic acid (EDTA)
- ☆ 1.5 M Sodium carbonate
- ☆ 2.25 mM NBT
- ☆ Riboflavin

Procedure (by Dhindsa *et al.*, 1981)

1. Superoxide dismutase (SOD) activity is measured immediately in fresh extract. Enzyme extract for SOD is prepared by grinding 0.1 g of leaf samples with 1 mL of extraction buffer (100 mM Potassium phosphate buffer (pH 7.5 + 0.1 mL of 0.5 mM EDTA).
2. The homogenate is centrifuged at 10,000 rpm for 10 min at 4°C and the supernatant is used as enzyme source.
3. The reaction mixture contains 1.5 mL of 100 mM phosphate buffer, 0.2 mL of 200 mM methionine, 0.1 mL of 3 mM EDTA, 0.1 mL of 1.5 M sodium carbonate, mL enzyme extract, 0.1 mL of 2.25 mM NBT and 0.9 mL distilled water.

4. The reaction is started by adding 0.1 mL of riboflavin and placing the tubes covered with black paper under two 15 W fluorescent lamps for 15 min.
5. A complete reaction mixture without enzyme, which gives the maximal colour, served as control. Switching off the lights and putting the tubes into dark stops the reaction. The non-irradiated complete reaction mixture serves as blank.
6. Absorbance of the samples is measured immediately after the reaction is stopped at 560 nm. The enzyme activity (grams per fresh weight) of a sample is determined from a standard curve obtained by using pure SOD.

Calculation

Per cent Inhibition of NBT reduction by SOD = control OD- treatment OD/ control x 100 =X per cent inhibition.

50 per cent inhibition is equal to 1 unit of enzyme.

Then X per cent is equal to 1/50 x X= Y unit.

Precautions

1. All buffers/solutions should be prepared using Millipore water and stored at 4°C until they are used and before the experiment, they must be brought to room temperature for optimal enzyme activity.
2. The crude leaf extract should be prepared fresh daily from leaf tissue and all assays must be conducted within a few hours.
3. The crude leaf extract should at all times be kept on ice.
4. Absorbance readings for SOD samples should be taken as soon as the assay is stopped, because the blue formazon will quickly start precipitating out, resulting in errors.
5. For all assays, care must be taken to avoid air bubbles forming in the cuvettes before making absorbance measurements.

6. CHITINASE CONTENT

Introduction

Chitinase occurs naturally in a wide variety of higher plants and catalyzes the hydrolysis of chitin, a linear polymer of β (1, 4)-N-acetyl glucosamine, found in fungal cell walls and insect cuticles. Chitinase from plant tissue has been reported to be antifungal and cause partial dissolution of hyphal walls. These results suggest a role for chitinase in plant defense. Some chitinases function as pathogenesis-related (PR) proteins. The induction of Systemic Acquired Resistance (SAR) and the consequent accumulation of chitinase in plant tissues have been reported in response to plant attack by fungi, bacteria, viruses, nematodes and insects. The enzyme can also be induced by wounding, heat shock and the application of plant disease resistance activators or ethylene. Increase in chitinase activity has been correlated with an increase in disease resistance observed in plants expressing induced resistance. Chitinase has been used as a biochemical marker for SAR, and in experiments of plant defense responses and microbe interactions.

Materials Required

1. Infected and control leaf samples
2. Pre-chilled mortar and pestle
3. Centrifuge
4. Water bath
5. Sodium citrate buffer: 0.1 M
6. Bovine Serum Albumin (BSA): 7.5 mg

Procedure (by Giri *et al.*, 1998)

1. For enzyme extraction 0.5 g of infected and control leaf samples are weighed separately and macerated with 2 mL of 0.1 M sodium citrate buffer in pre-cooled mortar and pestle.
2. The homogenate is centrifuged at 10,000 rpm for 10 min at 10°C and the supernatant is used as crude source of chitinase.
3. For determination of chitinase activity, 0.5 mL of supernatant is added to 2 mL of chitin suspension containing 7.5 mg of BSA and is incubated in water bath at 37°C for 3 h.
4. From that, an aliquot of 0.1 mL is taken for the estimation of N-acetyl glucosamine as per the method of Nelson-Smogyi.
5. The chitinase activity is expressed in terms of μg N-acetyl glucosamine released per min per mg of protein.

Precaution

1. The crude leaf extract should be prepared fresh daily from leaf tissue, kept on ice and all assays must be conducted within a few hours.
2. Proper care should be taken while handling the chemicals.

7. CATALASE (CAT)

Introduction

Catalase is an important enzyme that acts to dissociate hydrogen peroxide (H_2O_2) into molecular oxygen (O_2) and water (H_2O). It is an intracellular enzyme that has been discovered in most facultative anaerobes and all aerobic bacteria, but is not present in obligate anaerobes. Catalase is the second most abundant enzymatic antioxidant (after superoxide dismutase), which attenuates the levels of reactive oxygen species that ubiquitously accompany pathological disorders.

The principally common method for measuring catalase activity is the UV spectrophotometric method, which depends on monitoring the change of 240 nm absorbance at high levels of hydrogen peroxide solution (≥30 mM). High levels of hydrogen peroxide (H_2O_2) immediately lead to inhibition of the catalase enzyme by altering its active site structure. Hydrogen peroxide acts to oxidize cobalt (II) to cobalt (III) in the presence of bicarbonate ions; this process ends with the production of a carbonato-cobaltate (III) complex ([Co $(CO_3)_3$] Co).

Materials Required

1. Infected and control leaf samples
2. Glass homogenizing tube
3. Centrifuge
4. Water bath
5. UV-visible spectrophotometer

Chemicals Required

1. 0.2 M H_2O_2
2. 0.01 M phosphate buffer, pH 7.0
   a. Solution A – 0.01M Sodium dihydrogen phosphate solution
   b. Solution B – 0.01M Disodium hydrogen phosphate solution

   Mix 39 mL of solution A and 61 mL of solution B and make up the total volume to 100 mL with distilled water.
3. Stock dichromate-acetic acid solution: Mixed a 5 per cent potassium dichromate is mixed with glacial acetic acid (1:3 ratio).
4. Working dichromate-acetic acid solution: The stock solution was diluted in 1:5 ratios with water to make the working dichromate-acetic acid solution.

Procedure

1. Tissues are homogenized in the ice cold 10 volume of isotonic buffer in a glass homogenizing tube. The homogenate is centrifuged for 5-10 min at 700g to remove the debris.

2. The reaction mixture contains 0.5 mL of 0.2 M H_2O_2, 1 mL of 0.01 M phosphate buffer and 0.4 mL of water. To initiate the reaction 0.2 mL of tissue homogenate is added. 2 mL of the dichromate-acetic acid reagent is added after 0, 30, 60 and 90 seconds of incubation.
3. To the control tube the enzyme is added after the addition of the acid reagent. The test tubes are kept in boiling water bath (60°C) for 10 min and then cooled.
4. The colour developed is read at 620 nm against a reagent blank in UV-visible spectrophotometer.
5. Standards of H_2O_2 in the range of 2-10 µM are taken and preceded as test with blank containing reagent alone.
6. The activity of catalase is expressed as µmole of H_2O_2 consumed/min/mg protein.

Calculation

CAT activity is determined by measuring the decrease in absorbance as a result of degradation of H_2O_2 ($\varepsilon = 39.4\ mM^{-1}\ cm^{-1}$), which is followed in the reaction

Precaution

1. All buffers/solutions should be prepared using Millipore water and stored at 4°Cuntil they are used and they are brought to room temperature before the experiment.
2. The crude leaf extract should be prepared fresh daily from leaf tissue, kept on ice and all assays must be conducted within a few hours.
3. For all assays, care must be taken to avoid air bubbles forming in the cuvettes before making absorbance measurements.
4. Proper care should be taken while handling the chemicals.

8. β-1,3 GLUCANASE

Introduction

β- 1,3 glucanase isozymes have been suggested as an important component of plant defense mechanisms against pathogens. The enzymes are induced by pathogens, digest fungal cell walls and release oligosaccharide elicitors that elicit the production of low molecular weight antifungal compounds.

Required Chemicals

1. 15 per cent polyacrylamide gel
2. 0.05M sodium acetate buffer
3. 75mL of 0.05M sodium acetate buffer
4. Methanol, water and acetic acid (5:5:2, v/v)
5. 0.3 g of 2,3,5- triphenyl Tetrazolium chloride 1.0M
6. NaOH

Procedure

1. Enzyme is extracted as per the standard procedure.
2. Native-PAGE is carried out on 15 per cent polyacrylamide gel for 4h at 8°C at 30mA constant current.
3. After electrophoresis, the gel is washed with water, incubated with 0.05M sodium acetate buffer for 5 min and then incubated at 40°C for 30 min in a mixture containing 75mL of 0.05M sodium acetate buffer and 1.0g of laminar is dissolved in 75mL of water by heating in a boiling water bath.
4. The gel is incubated in a mixture of methanol, water and acetic acid (5:5:2, v/v) for 5 min, washed with water and strained with 0.3g of 2,3,5-triphenyl Tetrazolium chloride in 200mL of 1.0M NaOH in a boiling water bath until red bands appear (about 10min)

9. PHENYLALANINE AMMONIA LYASE (PAL)

Introduction

PAL is the first committed enzyme of the cinnamate-related secondary metabolism in plants and catalyzes a deamination of phenylalanine, a carbon-carbon double bond is formed during the release of NH_3, yielding trans-cinnamic acid. PAL activity is measured as the rate of conversion of L-phenylalanine to trans-cinnamic acid at 290 nm. PAL activity was first measured in protein extracts from *Sorghum bicolor* by Koukol and Conn in 1961. This is one of the most intensively studied enzymes of plant secondary metabolism. The formation of phenylpropanoid phytolalexins after fungal infection involves a very rapid induction of PAL. In plants it is a key biosynthetic enzyme that catalyzes the first step in the synthesis of a variety of polyphenyl compounds and is mainly involved in defense mechanisms.

$$\text{L-Phenylalanine} \xrightarrow{\text{PAL}} \text{trans-Cinnamate} + NH_3$$

Required Chemicals

- ☆ 0.1 M potassium phosphate buffer
- ☆ Mercaptoethanol
- ☆ Borate buffer
- ☆ 0.1 M L- Phenylalanine (pH 8.7)
- ☆ 1 M TCA

Procedure (by Dickerson *et al.*, 1984)

1. Enzyme extract for PAL is prepared by grinding 0.5 g of leaf samples by adding5 mL of 0.1 M potassium phosphate buffer and 20 µl mercaptoethanol. Samples are centrifuged in 12,000 rpm for 20 mins in 4°C and the supernatant is used as enzyme extract.
2. Reaction mixture contains 0.2 mL enzyme extract, 0.5 mL borate buffer, 1.3 mL distilled water and 1 mL of 0.1 M L- Phenylalanine (pH 8.7). These are incubated for 30-60 mins at 32°C.
3. After this, 0.5 mL of 1 M TCA is added to the mixture and absorbance is taken at 290 nm.
4. Enzyme activity is expressed on a fresh weight basis of amount of transcinnamic acid (nmol min^{-1} g^{-1}) synthesis.

Calculation

$$\text{Units/mL enzyme} = \frac{(5\ \text{A290nm/min Test} - \Delta\ \text{A290nm/min Blank})\ (3)(\text{df})}{(19.73)(0.1)}$$

where,

3 = Total volume (in mL) of assay

df = Dilution factor

19.73 = Millimolar extinction coefficient of trans-cinnamate at 290nm

$$\text{Units/mg solid} = \frac{\text{units/mL enzyme}}{\text{mg solid/mL enzyme}}$$

$$\text{Units/mg protein} = \frac{\text{units/mL enzyme}}{\text{mg protein/mL enzyme}}$$

Precaution

1. The crude leaf extract should be prepared fresh daily from leaf tissue.
2. Care must be taken to avoid air bubbles forming in the cuvettes before making absorbance measurements.

10. CHLOROPHYLL CONTENT

Principle

Chlorophyll is an antioxidant compounds which are present and stored in the chloroplast of green leaf plants and mainly it is present in the green area of leaves, stems, flowers and roots. Leaf chlorophyll concentration is an important parameter that is regularly measured as an indicator of chloroplast content, photosynthetic mechanism and of plant metabolism. Chlorophyll content will be changed with the change of external environment, which could further lead to the photosynthetic capacity change. Therefore, chlorophyll content could be used as an important diagnostic indicator for plant growth study. Chlorophyll a is the primary photosynthetic pigment in plants which helps to produce energy in plant and its concentration is 2-3 times higher than that of secondary chlorophyll b in plants. Variation in leaf chlorophyll content can provide information about the physiological condition of a leaf or plant. The assessments of chlorophyll contents are based on the extraction of chlorophyll with solvents from the destructive leaf followed by spectrophotometric determination.

Materials Required

- Mortar and pestle
- 100 mL volumetric flask
- Refrigerator
- Centrifuge
- UV- visible spectrophotometer
- 80 per cent acetone
- $MgCO_3$
- Leaf sample

Procedure (by Kamble *et al.*, 2015)

1. One gram of leaf sample is finely cut and gently mixed with a clean pestle and mortar. To this homogenized leaf material, 20 mL of 80 per cent acetone and 0.5 gm $MgCO_3$ powder is added.
2. The materials are further ground gently. The sample is then put into a refrigerator at 4°C for 4 hours. Thereafter, the sample is centrifuged at 500 rpm for 5 minutes.
3. The supernatant is transferred to 100 mL volumetric flask. The final volume is made up to 100 mL with addition of 80 per cent acetone.
4. The colour absorbance of the solution is estimated by UV- visible spectrophotometer using 645 and 663 nm wavelength against the solvent.
5. Acetone (80 per cent) is used as blank.

Calculation

Chlorophyll a (mg/g) = (12.7 × A663) – (2.59 × A645)

Chlorophyll b (mg/g) = (22.9 × A645) – (4.7 × A663)

Chlorophyll total (mg/g) = (8.2 × A663) + (20.2 × A645)

Precaution

1. Care must be taken to avoid air bubbles forming in the cuvettes before making absorbance measurements.
2. Absorbance must be measured carefully.

11. PROLINE CONTENT

Introduction

Accumulation of proline in higher plants is an indication of disturbed physiological condition, triggered by biotic or abiotic stress condition. Free proline content can increase upon exposure of plants to drought, salinity, cold, heavy metals, or certain pathogens. Determination of free proline levels is a useful assay to monitor physiological status and to assess stress tolerance of higher plants. Three methods suitable for determination of free proline content are described below. The isatin paper assay is simple and is suitable to assay proline content in large number of samples. The colorimetric measurement is quantitative and provides reliable data about proline content. The HPLC- based amino acid analysis can be employed when concentration of all amino acids has to be compared.

Proline Determination with Isatin Paper Assay

Isatin (1H-indole-2, 3-dione) is employed as a visualizing agent giving different colors with amino acids in thin layer and paper chromatography. Isatin is a highly specific color reagent for the proline, forming a blue derivative, pyrrole blue. Intensity of the blue color correlates with proline concentration. The characteristic blue coloration can therefore be used for rapid determination of free proline in biological samples. Other amino acids present in biological samples usually do not interfere with the result of free proline quantitation. This method is simple, fast, and is suitable to estimate proline content in large number of samples. Although not frequently used, utility of proline determination with isatin has been demonstrated in several papers.

Proline Determination with Colorimetric Assay

The ninhydrin-based colorimetric assay has been improved by Bates *et al.* (1973) and became the standard method for many laboratories to determine proline content. At acidic pH, ninhydrin can form a red product with proline and ornithine which can be used for the estimation of the concentration of these amino acids in pure solution. This assay is simple, reliable, quantitative, does not need sophisticated instrumentation or expensive reagents, and has been tested for numerous plants.

Amino Acid Determination with HPLC

The amino acid content of biological samples can be determined by reverse-phase, high- performance liquid chromatography (HPLC) combined with detection by UV and/or fluorescent and/or mass spectroscopy. The HPLC method is more sensitive for detecting a large range of amino acids than are methods using fluorescamine and ninhydrin. This method is suitable for the detection of amino acids such as proline and hydroxyproline. In that case, chloramine-T and sodium borohydride should be added to samples in order to produce proline and hydroxyproline derivatives that can react with OPA reagent. This method is useful to compare all amino acid content but needs quite sophisticated instrumentation.

Proline Determination with Colorimetric Assay

Materials Required

- Sulfosalicylic acid (3 per cent): Dissolve 3 g 5-sulfosalicylic acid (2-hydroxy-5- sulfobenzoic acid) in 80 mL distilled water and make up the volume to 100 mL. Solution can be stored at room temperature for weeks.
- Acidic ninhydrin: 1.25 g ninhydrin (1,2,3-indantrione monohydrate), 30 mL glacial acetic acid, 20 mL of 6 M orthophosphoric acid, dissolve by vortexing and gentle warming. Solution can be stored at 4°C for up to 1 week.
- Phosphoric acid solution (aqueous solution of 85 per cent orthophosphoric acid).
- Glacial acetic acid
- Toluene
- L-Proline
- Microcentrifuge
- Spectrophotometer
- Quartz cuvettes

Procedure

1. The samples are harvested, their fresh weights are measured, and approximately 100 mg is used for a reaction. Alternatively, the samples are snap freezed in liquid nitrogen. When necessary, samples can be stored at -80°C.
2. 3 per cent sulfosalicylic acid is added (5 µL/mg fresh weight) and the plant material is ground. The tubes are kept on ice until finishing with all samples.
3. Then the samples are centrifuged for 5 min at room temperature using benchtop centrifuge with maximum speed.
4. The reaction mixture is prepared in a separate tube: 100 µL of 3 per cent sulfosalicylic acid, 200 µL glacial acetic acid and 200 µL acidic ninhydrin.
5. About 100 µL of the supernatant of the plant extract is taken and the tubes are mixed well. To avoid high pressure and accidental opening of the tubes in subsequent reaction, you can puncture the lid of the microcentrifuge tube with a needle.
6. The tubes are incubated at 96°C for 60 min.
7. The reaction is terminated on ice.
8. The samples are extracted with toluene: 1 mL toluene is added to the reaction mixture, the samples are vortexed for 20 s, and it is left on the bench for 5 min to allow the separation of the organic and water phases. Gloves should be used to handle the tubes during extraction.

9. The chromophore containing toluene is removed into a fresh tube.
10. The absorbance is measured at 520 nm using toluene as reference. The proline concentration can be determined using a standard concentration curve and calculated on fresh weight basis (usually expressed as microgram per gram FW or micromole per gram FW).

Precaution

1. The fresh weight of samples should be determined carefully. If the samples are wet, drain off as much excess liquid as you can but do not dry plant samples.
2. Rubber gloves must be used when handling the reaction tubes. Ninhydrin is very reactive.
3. As toluene damages the plastic cuvettes, quartz glass cuvettes should be used to measure absorbance with spectrophotometer.
4. Toluene is harmful and highly flammable; handling of toluene should be done in chemical hood.

12. LIGNIN

Introduction

Lignin is one of the most abundant aromatic biopolymers in nature and accounts for nearly 30 per cent of the plants. Its deposition in the cell wall plays vital roles for plant development, such as providing rigidity and strength to the cell wall to give structural support for the plant cell wall. Lignin exhibits hydrophobicity in favor of the transport of water and solutes in the vascular system. It also protects the cell against pathogens.

The acetyl bromide, thioglycolic acid, and Klason are the more common methods used to quantify lignin. The two first methods are based on solubilization of lignin and determination of absorbance values at 280 nm. Thioglycolic acid method is based in the formation of thioethers of benzyl alcohol groups found into the lignin, resulting in solubilization of this polymer under alkaline conditions.

Required Chemicals

- ☆ Methanol
- ☆ 0.2 and 0.5 N NaOH
- ☆ 0.1M and 2M HCl
- ☆ Thioglycolic acid

Procedure

1. One gram of leaf tissues is extracted in two changes of 2 mL of methanol over the course of 24 h, the supernatant is discarded, centrifuged at 10,000 rpm for 10 min, and the solid are dried in an oven at 37°C for 48 h and their dry weights are determined.
2. The samples are then ground in 2 mL of 0.2 M NaOH and extracted for 24 h at room temperature to hydrolyze wall-bound phenolic acids.
3. The incubation mixture is then neutralized with 0.5 mL of 2M HCl and the residue is collected by centrifugation. Residues are washed with two changes of distilled water and then resuspended in 5 mL of methanol and the solids are collected by centrifugation.
4. The air-dried pellet is resuspended in 5 mL of 2M HCl to which 0.5 mL of thioglycolic acid is added. The tubes are sealed and placed in water bath at 95°C for 4 h, then cooled on ice and the solids are collected by centrifugation.
5. The solids are washed twice with 2 mL of distilled water each time, followed by centrifugation. The solids are then suspended in 0.5 M NaOH (5 mL) overnight at 4°C and again collected by centrifugation and washed twice with 2 mL H_2O.

6. The water washes are added to the NaOH supernatant, acidified with 1 mL of concentrated HCl, and the solution is placed at 4ºC for 4 h to precipitate the lignothioglycolic acid (LTGA).
7. The precipitate is collected by centrifugation and washed twice with 0.1 M HCl. The final pellet is dissolved in 0.5 M NaOH (3 mL) and insoluble material is removed by centrifugation.
8. The absorbance of 1 mL solution of LTGA is determined at 280 nm and the lignin content is expressed as absorbance at 280 nm mL^{-1} g^{-1} dry weight.

Precaution

1. Proper care should be taken while handling he chemicals and glass wares.
2. Care must be taken to avoid air bubbles forming in the cuvettes before making absorbance measurements.

13. CELL MEMBRANE STABILITY INDEX

Introduction

Environment stress causes cellular membrane modification in plants, which results in its perturbed function or total disfunction. The extract functional and structural modification caused by stress is not fully resolved. However, the cellular membrane disfunction due to stress is well expressed in increased permeability which can be leakage of ions outside and that can be measured by the efflux of electrolytes. Hence, the estimation of membrane disfunction under stress by measuring cellular electrolytes leakage from affected leaf tissue into an aqueous medium is finding a growing use as a measure of MSI and as a screen for stress resistance.

Materials Required

- ✰ Leaf sample
- ✰ Test tubes
- ✰ Distilled de-ionize water
- ✰ Water bath
- ✰ Electrical conductivity meter

Procedure

1. CMSI is assayed by estimating the electrolyte leakage from fresh leaf tissue into the distilled water using the method described by Sairam and Srivastava (2002).
2. One hundred milligram of leaf tissues from different samples are cut into 5mm length and placed in test tube containing 10 mL of distilled de-ionized water in two sets.
3. One set of leaf samples is placed in a water bath maintained at constant temperature at 40°C. After 2 h, conductivity (C_1) is measured using electrical conductivity meter.
4. The second set is placed in a boiling water bath at 100°C for 20 minutes to completely kill the tissue and release all electrolytes. Samples are cooled to 25°C and conductivity is measured (C_2).
5. CMSI is calculated using the formula of Premachandran *et al.* (1989).

Calculation

$$\text{CMSI} = 1 - C_1/C_2 \times 100$$

where,

C_1 = Conductivity after putting the sample at 40°C

C_2 = Conductivity after putting the sample at 100°C

14. MALONDIALDEHYDE (MDA) CONTENT

Principle

Malondialdehyde is the three carbon dialdehyde that is produced in plant species as the end product of polyunsaturated lipid peroxidation. This is most abundant aldehyde resulting from lipid peroxidation. Lipid peroxidation is a complex processing in which polyunsaturated fatty acids (PUFAs) are subjected to attack via oxygen-derived free radicals resulting in the formation of lipid hydroperoxides. In biological tissues, these lipid hydroperoxides are broken down from a variety of products such as aldehydes and ketones. MDA is a broadly used marker of oxidative lipid damage whose concentration varies in response to biotic and abiotic stress. For the determination of MDA, TBA assay is one of the widely used assays.

Materials Required

- ☆ Fresh leaf sample
- ☆ Thiobarbituric acid (TBA): 0.5 per cent
- ☆ Trichloroacetic acid (TCA): 0.1 per cent
- ☆ Ice bath
- ☆ Centrifuge
- ☆ UV visible spectrophotometer

Procedure

1. Lipid peroxidation is measured in terms of malondialdehyde (MDA) in the leaf based on reaction with thiobarbituric acid (TBA) following the procedure described by Cakmak and Horst (1991).
2. Fresh samples of leaf (500 mg) are homogenized in 10 mL of 0.1 per cent trichloroacetic acid (TCA). Then homogenates are centrifuged at 15,000 rpm for 5 min.
3. Then 2 mL aliquot of supernatant is taken and 0.5 per cent thiobarbituric acid (TBA) in 20 per cent TCA is beaded into it.
4. The mixture is heated at 95°C for 30 minutes and then quickly cooled in ice bath followed by centrifugation at 1000 rpm for 10 minutes.
5. Suspended turbidity is removed.
6. The absorbance of supernatant is recorded at 532 nm wavelength on UV visible spectrophotometer.
7. The value for nonspecific absorption at 600 nm is subtracted and the MDA content is calculated using its absorption coefficient of 155 $mmol^{-1}$ cm^{-1}.

Calculation

Absorbance due to MDA-TBA abduct = Absorbance at 532) - (Absorbance at 600)

Extinction coefficient of this MDA-TBA abduct at 532 nm is 155 $mM^{-1}cm^{-1}$.

Concentration of MDA (mM) = (A532 - A600)/155

Volume of reaction mixture is 2 mL

Suppose,

Concentration of MDA (mM) = (A532 - A600)/155 = X

Concentration of MDA (microM)= X x 1000 = Y

MDA present in 2 mL will be Y x 2 = 2Y nmoles

15. ASCORBATE PEROXIDASE (APX)

Introduction

Under biotic and abiotic stress conditions, plants generate numerous reactive oxygen species (ROS), including superoxide radical, hydrogen peroxide (H_2O_2), and lipid peroxides. Excessive accumulation of ROS damages plant cells via lipid peroxidation and protein oxidation. Many enzymatic and non-enzymatic antioxidants can reduce ROS levels to maintain cellular redox balance. Ascorbate peroxidase (APX) is a plant antioxidant enzyme that is important in the synthesis of ascorbate (AsA), a natural plant antioxidant capable of scavenging free radicals generated by different stresses in plants and scavenging of excess hydrogen peroxide. APx are distributed in stroma, thylakoids, microbodies, cytosol, and mitochondria.

APx can catalyze the reaction between ascorbic acid (AsA) and hydrogen peroxide (H_2O_2) and AsA can be oxidized to Monodehydroascorbic acid (MDASA). The absorbance of solution at 290 nm will decline as the oxidation of ascorbate.

Materials Required

- ☆ Infected and control leaf sample
- ☆ Mortar and pestle
- ☆ Centrifuge
- ☆ UV-visible spectrophotometer

Chemicals Required

- ☆ 100 mM Potassium phosphate buffer (pH 7.5)
- ☆ Ascorbic acid
- ☆ EDTA (Ethylenediaminetetraacetic acid)
- ☆ Hydrogen peroxide (H_2O_2)

Procedure

1. Enzyme extract for APx is prepared by grinding 0.5 g of leaf sample with 2 mL of 100 mM Potassium phosphate buffer (pH 7.5). The homogenate is centrifuged at 15,000 rpm for 15 min at 4°C and the supernatant is used as enzyme source.
2. The reaction mixture contains 2.3 mL phosphate buffer, 0.2 mL ascorbic acid, 0.2 mL EDTA, 50 µl enzyme extract, 50 µl hydrogen peroxide (H_2O_2) and 0.3 mL distilled water.
3. The reaction is started with addition of 0.2 mL of hydrogen peroxide. Decrease in absorbance after 30 sec is measured at 290 nm in UV-visible spectrophotometer.

4. The activity is determined using molar extinction coefficient U = 2.8 mM^{-1} cm^{-1} and expressed in terms of millimole of ascorbate per minute per gram fresh weight.

Calculation

Enzyme activity (Units/L) = (Δ Abs × Total assay volume)/(Δt x ε x l x Enzyme sample volume).

where,

Δt = Time of incubation (min)

ΔAbs = change in absorbance

ε = Extinction coefficient of substrates (M^{-1} cm^{-1})

l = Cuvette diameter (1cm)

Enzyme activity (Unit) is defined as the amount of enzyme that oxidizes 1μmol of substrate/min.

Precautions

1. All buffers/solutions should be prepared using Millipore water and stored at 4°C until they are used and they are brought to room temperature before the experiment.
2. The crude leaf extract should be prepared fresh daily from leaf tissue, kept on ice and all assays must be conducted within a few hours.
3. Ascorbic acid in the APx assay should always be kept on ice.
4. For all assays, care must be taken to avoid air bubbles forming in the cuvettes before making absorbance measurements.

16. GLUTATHIONE PEROXIDASE (GPX)

Introduction

Free radicals are continuously produced *in vivo* and there are a number of protective antioxidant enzymes for dealing with these toxic compounds. Glutathione peroxidase is one of the free radical scavenger enzymes, which is reduced progressively due to exposure to pathogen infection (Rotruck *et al.*, 1973).

Materials

- ☆ 0.4 M Sodium phosphate buffer (pH 7.0)
- ☆ 10 mM sodium oxide
- ☆ 4 mM Glutathione, reduced
- ☆ 2.5 mM Hydrogen peroxide
- ☆ 10 per cent Trichloroacetic acid
- ☆ 0.3 M Disodium hydrogen phosphate buffer
- ☆ 5, 5′- Dithiobis 2- nitrobenzoic acid (Dissolve 40 mg in 100 mL of 1 per cent sodium citrate solution)
- ☆ Standard reduced glutathione – Prepare 20 per cent mg solution

Procedure

1. Prepare 10 per cent tissue homogenate in phosphate buffer (pH 7.0) and centrifuge at 10,000 rpm for 15 min at 4°C.
2. Take 0.4 mL buffer solution, 0.1 mL sodium azide, 0.2 mL of reduced glutathione, 0.5 mL of enzyme extract and 0.1 mL of hydrogen peroxide into a test tube.
3. Make up the final volume to 2.0 mL with sterile distilled water. Incubate the tube at 37°C for 30 min.
4. Add 0.5 mL of 10 per cent trichloroacetic acid to stop the reaction.
5. To determine the residual glutathione content, centrifuge the content, save the supernatant and then add 3.0 mL of disodium hydrogen phosphate and 1.0 mL of DTNB reagent.
6. Read color were developed and absorbance was recorded at 412 nm.
7. The activity is expressed as µg of glutathione consumed/minute/mg protein.

17. ASCORBIC ACID OXIDASE

Introduction

Ascorbic acid oxidase is widely found in plant tissues. The crucial role of this enzyme is to regulate the levels of oxidized and reduced glutathione and NADPH. The activity of this enzyme is induced during pathogen infection.

Materials

- Phosphate buffer 0.1 M (pH 5.6)
- Substrate solution- dissolve 8.8 mg ascorbic acid in 300 mL phosphate buffer

Procedure

1. Take 1 mg of plant tissue macerate in 5 mL of phosphate buffer. Centrifuge the homogenate at 5000 rpm for 15 min. Use the supernatant as enzyme source.
2. Pipette out 3 mL of substrate solution to each of sample and cuvettes.
3. Add 0.1 mL of enzyme extract to the cuvette.
4. Absorbance was recorded change at 265 nm in 30 second intervals for 5 min.

Calculation

From the linear phase of reaction, compute the change in absorbance per min. Activity may be expressed in enzyme units. One enzyme unit is equivalent to 0.405 µmole oxygen per min. Ascorbic acid requires only one atom of oxygen per mole oxidized.

18. TOTAL CARBOHYDRATE BY ANTHRONE METHOD

Principle

Carbohydrates are first hydrolyzed into simple sugars using dilute hydrochloric acid. In hot acidic medium glucose is dehydrated to hydroxylmethyle furfural. This compound form with anthrone a green-colored product with an absorption maximum at 630nm. Anthrone react with dextrins, monosaccharides, disaccharides, polysaccharides, starch, gums and glycosides. But they yield of colour where is to form carbohydrate to carbohydrate.

Material Required

- 2.5 N HCl
- Antrone reagent: Dissolve 200 mg anthrone in 100 ml of ice-cold 95 per cent H_2SO_4. Prepare fresh before use
- Standard glucose: Stock- Dissolve 100 mg in 100 ml water. Working standard -10 mL of the stock diluted 10 100 mL with distilled water. Store refrigerated after adding a few drops of toluene.

Procedure

1. Weight 100 mg of the sample into a boiling tube.
2. Hydrolyze by keeping it in boiling water bath for three hours with 5 ml. of 2.5 N HCl and cool to room temperature.
3. Neutralize is with solid sodium carbonate until the effervescence ceases.
4. Make the volume to 100 mL to centrifuge.
5. Collect the supernatant and take 0.5- and 1-mL aliquots for analysis.
6. Prepare the standard by taking 0, 0.2, 0.4, 0.6, 0.8 and 1 mL of the working standard '0' serves as blank.
7. Make up the volume to 1 mL in all the tubes including the sample tubes by adding distilled water.
8. Then add 4mL of anthrone reagent.
9. Heat for eight minutes in a boiling water bath.
10. Cool rapidly and read the green to dark green colour at 630nm.
11. Draw a standard graph by plotting concentration of the standard on the X-axis versus absorbance on the Y-axis.
12. From the graph calculate the amount of carbohydrate Present in the sample tube.

Observation

Green to dark green colour appeared observe at 630nm.

Calculation

Amount of carbohydrate Present in 100 mg of the sample

$$= \frac{\text{Mg of glucose}}{\text{Volume of the test sample} \times 100}$$

Precautions

1. Weight exactly 100 mg of glucose while making stock.
2. Keep test tubes for 10 minutes in water bath. Use a stop watch for this purpose.
3. Cool the solution rapidly after removing from water bath.

Chapter 2

Histochemical Analysis

1. Hydrogen Peroxide (H_2O_2) (Thordal-Christensen *et al.*, 1997)

Introduction

In this protocol, the *in-situ* detection of hydrogen peroxide (one of several reactive oxygen species) is described in potato leaves by staining with 3,3′-diaminobenzidine (DAB). DAB is oxidized by hydrogen peroxide in the presence of some haem-containing proteins, such as peroxidases, to generate a dark brown precipitate. This precipitate is exploited as a stain to detect the presence and distribution of hydrogen peroxide in plant cells. The protocol can be modified slightly to detect hydrogen peroxide in different types of plant tissue.

Materials

- ☆ Potato leaves
- ☆ Absolute ethanol
- ☆ Amber-colored bottle
- ☆ Glycerol
- ☆ Aluminium foil
- ☆ Test tubes
- ☆ Distilled water
- ☆ Measuring cylinder
- ☆ 3,3′-Diaminobenzidine (DAB)
- ☆ Bleaching solution
- ☆ Sodium phosphate
- ☆ Paper towel

Reagents

Preparation of DAB Solution

1. In 50 mL amber coloured bottle, add 50 mg DAB and 45 mL sterile distilled water for a final 1 mg mL^{-1} DAB solution.
2. Add 0.2 M HCl to reduce pH (3.0), and dissolve DAB.
3. Cover amber coloured bottle with aluminium foil.
4. Add 2.5 mL 50 mM Na_2HPO_4 to the stirring DAB solution.

Preparation of Sodium Phosphate (NaH_2PO_4) Buffer

1. Prepare 1 M Sodium phosphate monobasic (Na_2HPO_4) solution by dissolving 12 g Na_2HPO_4 in a final volume of 100 mL distilled water.
2. Prepare 1 M Sodium phosphate dibasic (Na_2HPO_4) solution by dissolving 14.2 g Na2HPO4 in a final volume of 100 mL distilled water.
3. Mix 16 mL 1 M Na_2HPO_4 and 84 mL 1 M Na_2HPO_4 solution and make the volume to 2L with distilled water for preparing 50 mM sodium phosphate buffer.

Preparation of Bleaching Solution

1. Ethanol: acetic acid: glycerol = 3:1:1

Procedure

1. Fresh potato leaves are randomly collected from each treatment after pathogen inoculation.
2. Collected leaf samples are washed with distilled water to remove other extraneous material associated with the leaf.
3. The potato leaves are placed in test tubes and immersed in DAB staining solution.
4. The tubes are covered with the help of aluminium foil and keep in dark conditions for 3 h.
5. The staining solution is drained off from the test tubes.
6. The chlorophyll is removed for clear visualization of the stain. This is done by immersing the leaves in bleaching solution and then leaves are incubated at 100°C for 10 min.
7. The leaves are then transferred on a paper towel saturated with 60 per cent glycerol.
8. Photographs of the stained leaves are taken against a contrast background for proper visualization.
9. Then it is observed under the light microscope. The presence of DAB, hydrogen peroxide (H_2O_2) is visualized as a reddish-brown color formation in the potato leaves (**Figure 1**).

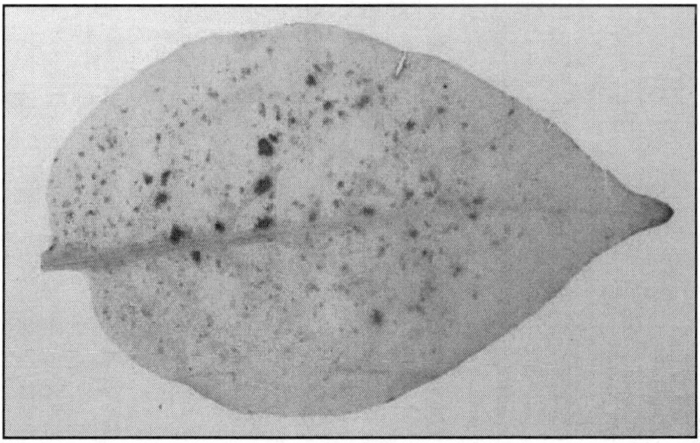

Figure 1: Detection of Hydrogen Peroxide Accumulation in Potato Leaves Exposed to Pathogen Stress for 5 Days.

2. SUPEROXIDE (O_2) (Kumar *et al.*, 2014)

Introduction

Plant cells continually produce reactive oxygen species (ROS) as a by-product of aerobic metabolism. Increased production of ROS occurs under unfavorable conditions imposed by various abiotic and biotic factors. Accumulation of ROS is damaging to various cellular components and macromolecules including plasma membrane, nucleic acids, and proteins and eventually leads to cell death. Superoxide (O_2^-) anion, most important ROS, in potato leaves by using Nitro tetrazolium Blue Chloride (NBT) as the chromogenic substrate. NBT reacts with O_2^- to form a dark blue insoluble formazan compound.

Materials

- Potato leaves
- Amber-colored bottle
- Distilled water
- Test tube
- Measuring cylinder
- Aluminium foil
- Absolute ethanol
- Glycerol
- Nitro tetrazolium Blue Chloride (NBT)
- Bleaching solution
- Sodium phosphate buffer
- Magnetic stirrer

Reagents

Preparation NBT Staining Solution

1. Dissolve 0.1 g NBT in 50 mM sodium phosphate buffer (pH 7.5) and make the volume to 50 mL to get a 0.2 per cent solution in an amber-colored bottle.
2. Mix the solution thoroughly using a magnetic stirrer.
3. NBT staining solution should be prepared freshly before use.

Preparation of Sodium Phosphate (NaH_2PO_4) Buffer

1. Prepare 1 M Sodium phosphate monobasic (NaH_2PO_4) solution by dissolving 12 g NaH_2PO_4 in a final volume of 100 mL distilled water.
2. Prepare 1 M Sodium phosphate dibasic (Na2HPO4) solution by dissolving 14.2 g Na2HPO4 in a final volume of 100 mL distilled water.

3. Mix 16 mL 1 M NaH_2PO_4 and 84 mL 1 M NaH_2PO_4 solution and make the volume to 2L with distilled water for preparing 50 mM sodium phosphate buffer.

Preparation of Bleaching Solution

1. Ethanol: acetic acid: glycerol = 3:1:1

Procedure

1. Potato leaves are sampled randomly after pathogen inoculation for each treatment.
2. Collected leaf samples are washed with distilled water to remove other extraneous materials associated with the leaf.
3. The potato leaves are placed in test tubes and immerse them in NBT staining solution.
4. The tubes are t hen covered with the help of aluminium foil and incubated at room temperature(25°C) in dark for 2 h.
5. Afterward, leaf sample are decolorised with the help of bleaching solution and heated at 100°C for 20 min.
6. Development of bluish spots shows the formazan deposits on the leaf surface expressed the visualization of the superoxide radicals (**Figure 2**).

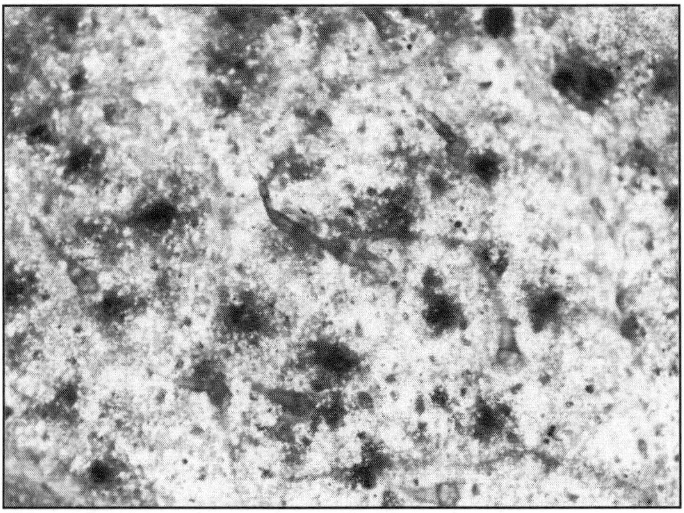

Figure 2: Detection of Superoxide Radical's Accumulation in Potato Leaves Tissues Exposed to Pathogen Stress for 5 Days.

3. CELL DEATH (Iannone *et al.*, 2015)

Introduction

Cell death is an important variable to studies host and pathogen interaction. The protocol is applied specifically to detect cell death on potato leaves during the course of infection with necrotrophic fungus *Alternaria solani*.

Materials

- ☆ Potato leaves
- ☆ Ethanol (95 per cent)
- ☆ Test tubes
- ☆ Measuring cylinder
- ☆ Evan's blue
- ☆ Aluminium foil
- ☆ Conical flask

Reagents

Preparation Evan's Blue Staining Solution

1. Weight 0.1 gm Evan's blue in 100 mL distilled water.
2. Evan's blue staining solution should be prepared freshly before use.

Procedure

1. Potato leaves are sampled randomly after pathogen inoculation for each treatment.
2. Collected leaf sample are washed with distilled water to remove other extraneous material associated with the leaf.
3. The potato leaves are placed in test tubes and immersed in 0.1 per cent Evan's blue staining solution.
4. The tubes are covered with the help of aluminium foil and incubated at room temperature(25°C) in dark for 20 min.
5. Then boiling of the leaf samples is done in 95 per cent ethanol for decolorising for 30 min.
6. Photographs of the stained leaves are taken against a contrast background for proper visualization.
7. The areas expressed necrotic lesions are observed as indigo blue spots color on the leaves (**Figure 3**).

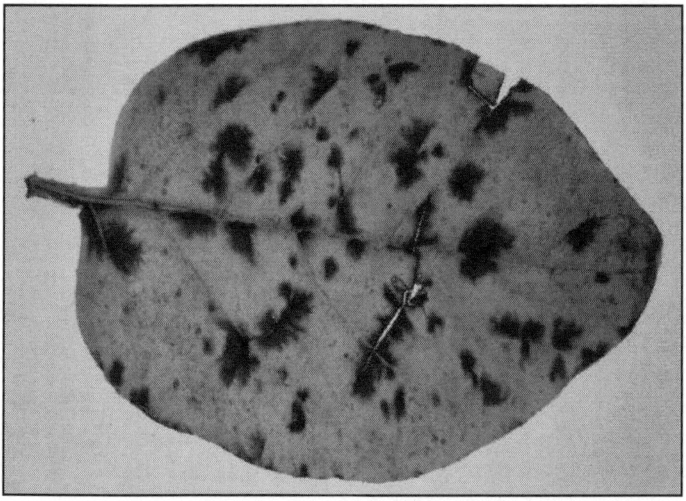

Figure 3: Detection of Cell Death Accumulation in Potato Leaves Exposed to Pathogen Stress for 5 Days.

4. LIPID PEROXIDATION (Airaki *et al.*, 2012)

Introduction

Oxidative stress induced by both biotic and abiotic stress factors leads to lipid peroxidation. The peroxidation of unsaturated lipids in membrane is the most apparent symptom of oxidative stress. Lipid peroxidation is a deleterious process in plants, which affects membrane properties, causes protein degradation and limits the capacity of ionic transport, ultimately triggering the cell death process. Malondialdehyde (MDA) content, which is a by-product of lipid peroxidation process is found to be enhanced in various crops on exposure to stress. The reactive oxygen species (ROS) associated with oxidative stress acts on membrane lipids to decrease membrane stability. Schiff's reagent reacts with aldehyde functional group of MDA to give the magenta coloration, thus acting as a key determinant of lipid peroxidation in plant tissues.

Materials

- ☆ Potato leaves
- ☆ Measuring cylinder
- ☆ Test tubes
- ☆ Distilled water
- ☆ Schiff's reagent
- ☆ Ethanol
- ☆ Glycerol
- ☆ Acetic acid
- ☆ Bleaching solution
- ☆ Conical flask
- ☆ Aluminium foil

Reagents

Preparation of Schiff's Staining Solution

1. Prepare 10 per cent (v/v) Schiff's reagent solution in distilled water and mixed well.

Preparation of Bleaching Solution

1. Ethanol: acetic acid: glycerol = 3:1:1

Procedure

1. Potato leaves are sampled randomly after pathogen inoculation for each treatment.
2. Collected leaf sample are washed with distilled water to remove other extraneous material associated with the leaf.

3. The potato leaves are placed in test tubes and immerse them in 10 per cent Schiff's staining solution.
4. The tubes are covered with the help of aluminium foil and incubated at room temperature (25°C) in dark for 1-2 h.
5. Afterward, leaf sample are decolorised with the help of bleaching solution and heated at 100°C for 40 min.
6. Photographs of the stained leaves are taken against a contrast background for proper visualization.
7. Development of pink color spot on the leaf surface expresses the lipid peroxidation (**Figure 4**).

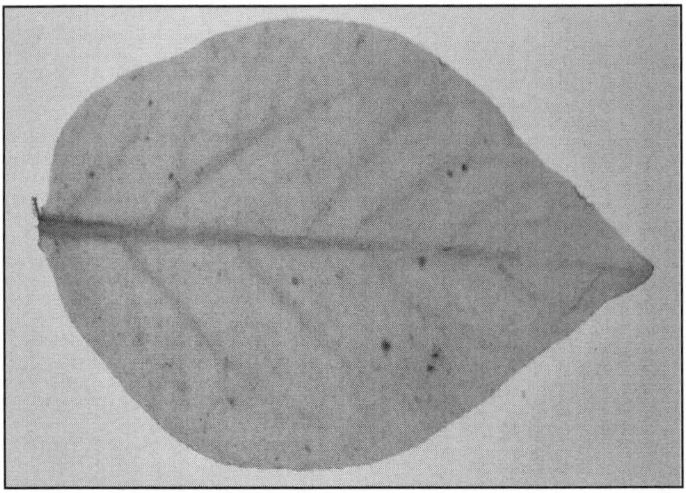

Figure 4: Detection of Lipid Peroxidation Accumulation in Potato Leaves Exposed to Pathogen Stress for 5 Days.

5. LIGNIN (Guo *et al.*, 2001)

Introduction

Lignin is a complex polymer of phenolic compounds (monolignins), which contributes to the rigidity of the plant cell wall. Lignification is essential for plant development; however, it is also one of the mechanisms of plant defense. Accumulation of lignin and the polymerization of monolignins at sides of pathogen attack protect the cell wall against cell wall-degrading enzymes and prevent therefore the pathogen's penetration. In addition to cross-linkage of phenolic compounds, this resistance mechanism includes also callose and cellulose appositions on the cell wall. This results in structures called papillae, which provide the necessary resistance to the mechanical pressure exercised by fungal appressorium. Lignin accumulation in cell walls is therefore a part of plant defense responses.

Materials

- ☆ Potato stems
- ☆ Microtome
- ☆ Ethanol (95 per cent)
- ☆ Phloroglucinol
- ☆ Distilled water
- ☆ Watch glass
- ☆ Slides and coverslips
- ☆ Brush
- ☆ Hydrochloric acid (HCl)
- ☆ Absorbent cotton
- ☆ Needle

Reagents

Preparation of Phloroglucinol Staining Solution

1. Weight 1 gm of phloroglucinol and dissolved in 95 per cent ethanol.
2. Mix the solution thoroughly using a magnetic stirrer.
3. Phloroglucinol staining solution should be prepared freshly before use.

Procedure

1. Potato stems are sampled randomly after pathogen inoculation for each treatment.
2. Collected stem samples are washed with distilled water to remove other extraneous material associated with the stem.
3. Transverse sections of the stem are cut with a freezing microtome.

4. Sliced stem is firstly dipped in distilled water for 3-4 min and then transferred with the help of needle in to 1 per cent phloroglucinol staining solution for 4 -5 min.
5. And then it is transferred in to 20 per cent concentrated HCl solution for few seconds and then transferred in to sterilized glass slide and covered with a coverslip and visualized under the light microscope (Nikon, Japan).
6. A positive area of lignin deposition is showed pink and red-violet color (**Figure 5**).

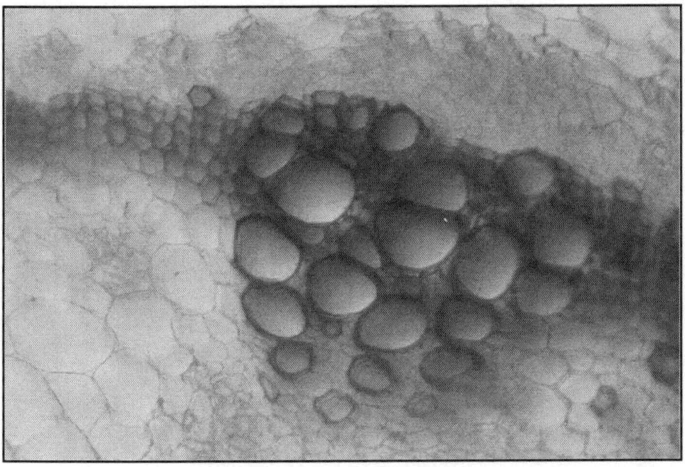

Figure 5: Detection of Lignin Accumulation in Potato Stems Exposed to Pathogen Stress for 5 Days.

6. PECTIN DEPOSITION (Singh *et al.*, 2020)

Materials

- ☆ Potato roots
- ☆ Ruthenium red
- ☆ Distilled water
- ☆ Glycerol
- ☆ Slides and coverslips

Reagents

Preparation of Ruthenium Red Staining Solution

1. Weight 0.2 mg ruthenium red and dissolved in 100 mL distilled water.
2. Mix the solution thoroughly using a magnetic stirrer.
3. Ruthenium red staining solution should be prepared freshly before use.

Procedure

1. Potato roots are sampled randomly after pathogen inoculation for each treatment.
2. Collected roots sample are washed with distilled water to remove other extraneous material associated with the root.
3. Transverse sections of the root are cut with a freezing microtome.
4. Sections are placed in 0.2 per cent ruthenium red staining solution for 30-35 min.
5. Sections are washed slowly to remove excess stain and mounted in glycerol and visualized under light microscope.
6. Development of pink color after staining shows presence of pectin deposition in root cells (**Figure 6**).

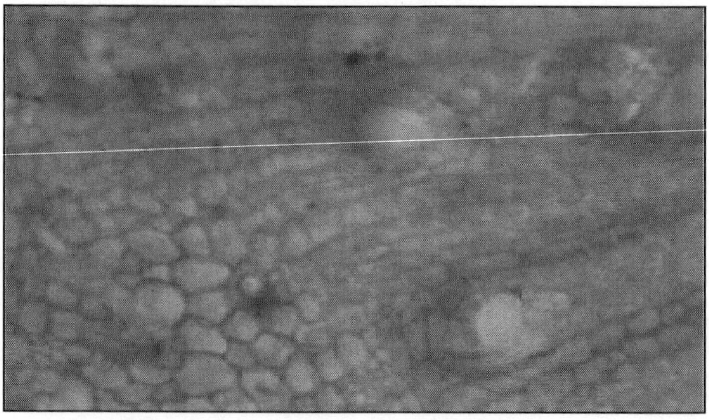

Figure 6: Detection of Pectin Deposition in Potato Stems Exposed to Pathogen Stress for 5 Days.

Chapter 3

Molecular Techniques Used in Identification and Characterization of Bacteria and Fungi

(Prassan *et al.*, 2019)

Introduction

Some time ago, the identification of bacteria and fungi based on morphological and culture characteristics are not provides accurate results require specific expertise and whole process is also time consuming. However, molecular tools offer the advantage to accurate identification of the microorganism and can effectively complement the phenotype-based identification. For identification of microorganism usually ribosomal DNA or spacer regions are used. However, such techniques have limited resolution of taxonomic delineation and are rarely used for differentiation beyond species level, molecular markers, which have wider genomic distribution are more effective in such cases. In a nut shell, at the present scenario molecular identification or characterization of microorganism is almost inevitable to establish any scientific advancement in microbiological research.

1. Molecular Identification and Characterization of Bacteria

I. 16S *r*RNA Gene-Based Identification

During the past decade ribosomal RNA based identification and detection methods have become techniques routinely used in all disciplines of microbiology. As molecular chronometers these molecules have preserved their evolutionary history. Highly conserved portions carry the information on early evolutionary evens and more recent changes are documented with in less conserved positions or stretches. The degree of divergence of present day rRNA sequences gives an estimate of their phylogenetic distances.

A. Reagents

- Taq DNA Polymerase (5 units/µl)
- 10 X Taq DNA polymerase buffer (0.1M Tris,0.5M KCl 1.5 mM MgC12,0.1 per cent gelatin)
- Nucleotide mix 10mM (dATP, dCTP, dGTP, dTTP)
- Lambda DNA (Hind III digest) molecular weight marker
- 100 bp ladder
- Electrophoresis buffer - TBE (5x) for 1 litre; 54 g Tris-HCl, 27.5 g Boric acid, 20 mL 0.5
- EDTA (pH 8.0); for run 0.5 x TBE is used.
- Primers

  Forward (5′AGAGTTTGATCCTGGCTCAG3′)

  Reverse (5′AAGGAGGTGATCCAGCCGCA 3′)

Procedure

Preparation of Template DNA from Gram Negative Bacteria

1. Grow bacteria at 30°C in 25 mL medium on a shaker, take 3 mL of broth culture and then centrifuge at 6000 rpm for 5 min.
2. Resuspend pellet in 0.4 mL TE, add 40µl of 10 per cent SDS, incubate at 37°C for 30 min in water bath.
3. Then add 0.4 mL phenol with well mixing, after mixing, centrifuge at 10,000 rpm for 3min.
4. Transfer the water phase to new Eppendorf tube and add 0.2 mL phenol and 0.2 mL chloroform and well mixing and centrifuge at 10,000 rpm for 3 min.
5. Transfer the water phase to new Eppendorf tube, add 0.4 mL chloroform and well mix, centrifuge at 10,000 rpm for 3 min.
6. Transfer the water phase to new Eppendorf tube and determine the total volume.
7. Add 0.1 volume 3 M NaAc (pH 5.2), well mix, then add 2 volume 96 per cent ethanol and incubate at -20°C for 1 h.
8. Centrifuge DNA at 10,000 rpm for 8 min, washed the pellet with 70 per cent ethanol and discard the supernatant.
9. To allow the dry pellet and dissolve in 40µl DEPC water and stored at 4°C for further used.

PCR Reaction Mixture

Component	*Volume*
Deionized water	78.0μl
10X Taq pol buffer	10.0μl
Taq polymerase	(1.0U)1.0μl
dNTPs (2mM)	6.0μl
Primer PA (100ng)	1.5μl
Primer PH (100ng)	1.5μl
Target DNA (50ng)	1.0μl

B. Preparation of the PCR Amplifications

1. Add the reagents in the required volumes and negative control without template DNA has to be maintained simultaneously to check for possible contamination.
2. If you are using a thermal cycler with a heated lid, the tubes are ready to go into the thermal cycler.
3. DNA amplification will be done in Thermal Cycler and amplified DNA will be examined by horizontal electrophoresis in 0.8 per cent agarose

C. Preparation of Agarose Gel

1. Prepare 0.8 per cent agarose solution and melt it in boiling water bath.
2. Cool the solution to 60°C and place the comb 0.5 - 10 mm above the plate so that a complete well is formed and then pour the agarose solution and let it solidify.
3. After the gel is completely set, carefully remove the comb and tape and mount the gel in the electrophoresis tank.
4. Add electrophoresis buffer to cover the gel to a depth of about 1mm and then add ethidium bromide (0.5 mg/mL) to the buffer.
5. Mix 20 μl sample of DNA with 8 ul of gel loading buffer (0.25 per cent bromophenol blue, 40 per cent (w/v) sucrose in water).
6. Load the mixture into the slots of the submerged gel using a disposable micropipette.
7. Load also Sul of lambda DNA Hind III digest marker or Sul of 100 bp ladders.
8. Close the lid of the gel tank and attach the electrical leads so that the DNA will migrate toward-the anode and gel observed under UV light.

D.

The amplicons are then purified and subjected to sequencing. The sequenced amplicons are then subjected to BLAST against a reference database (use EzTaxon) to determine the identity.

II. Amplified Ribosomal DNA Restriction Analyses (ARDRA)/16S *r*RNA PCR-RFLP

Introduction

This can be used as basic fingerprinting method or for primary analyses of diversity. Differentiations are made on the basis of restriction patterns obtained from different microbial species/strains.

Reagent(s)

- ☆ Lambda DNA (Hind III digest)
- ☆ Ladder (100 bp)
- ☆ 5X TBE buffer (548 Tris base, 20mL of 0.5 M EDTA pH 8.0 and 27.5 g of boricacid) and dissolved in one litre of water
- ☆ Ethidium bromide (5mg mL^{-1})
- ☆ Loading dye (6x) (0.25 per cent Bromophenol blue in 40 per cent sucrose w/v).
- ☆ Restriction endonucleases (*Alu* 1, *Bsu* R1, *Msp* 1 and *Rsa* 1) along with the buffer(10x)

Procedure

Step 1. Amplification of 16S rRNA Gene Employing PCR

- ☆ Primers

  Forward (5′AGAGTTTGATCCTGGCTCAG3′)

  Reverse (5′AAGGAGGTGATCCAGCCGCA 3′)

PCR Mix (25 μl total reaction)

Component	*Required Volume*
Nuclease free water	17.65 µl
DNA template	2.0 µl
Taq polymerase	1.0 U
Forward primer	0.6 µl
Reverse primer	0.6 µl
dNTPs (25 mM)	1.5 µl
10 X PCR buffer with $MgCl_2$	2.5 µl

PCR for the amplification of the 16S rDNA gene

Step1 = 95°C for 1 min	(Denaturation step)
Step 2 = 50°C for 30 Sec	(Primer Annealing Step)
Step 3 = 72°C for 1 min 20 Sec	(Elongation step)
Step-4 = 34 cycle	(Repeat Step 1 to 3)

Step5 = 72°C for 10 min (Final elongation step)

Hold at 4°Cfor 1h

Step 2. Restriction Digestion of 16S rRNA Gene by Restriction Endonuclease

1. The reaction mixture for restriction analysis should be prepared as follows

PCR product	12.3µ L
Restriction endonuclease (3U)	0.3µL
Recommended buffer (for restriction enzyme)	1.4 µL
Total volume	14.0µL

2. Incubation of the restriction endonuclease mixture keep the mixture tubes in water bath maintained at 37°C for 2 hrs.
3. Analyze the restricted DNA by horizontal electrophoresis in 3 per cent agarose gel added with the ethidium bromide. Carry out the electrophoresis at 100V for 2 h 30 min in 1X TAE buffer with standard gels (11x14 cm).
4. Visualize the gels under gel documentation and record your observations.

2. Molecular Identification of Fungi by Using ITS Region Sequencing

Introduction

DNA extraction from fungal mycelium can vary depending on the material used. Essentially any mechanical means of breaking down the cell wall and membranes to allow access to nuclear material, without its degradation is required. For this, usually an initial grinding stage with liquid nitrogen is employed to break down cell wall material and allow access to DNA while harmful cellular enzymes and chemicals remain inactivated. Once the mycelium has been sufficiently ground, it can then be resuspended in a suitable buffer, such as CTAB. In order to purify DNA, insoluble particulates are removed through centrifugation while soluble proteins and other material are separated through mixing with chloroform and centrifugation. DNA must then be precipitated from the aqueous phase and washed thoroughly to remove contaminating salts. The purified DNA is then resuspended and stored in TE buffer or sterile distilled water. This method has been shown to give intact genomic DNA from fungal mycelium.

Isolation of Genomic DNA from Fungal Mycelium

Reagents

- ☆ Lysis buffer: 50 mM Tris-HCl, 50 mM EDT, 3 per cent SDS, 1 per cent 2-mercaptoethanol (add just before use)
- ☆ Chloroform: phenol (1:1)

- ☆ SEVAG (chloroform: Isoamyl alcohol, 24:1)
- ☆ 3 M NaOAc (pH 8.0)
- ☆ Isopropanol
- ☆ Ethanol (100 per cent, -20°C)

Procedure

1. Fill a 1.5 mL eppendorf microcentrifuge tube 2/3 to the joint with ground lyophilized mycelium (60-100 mg dry, or 0.5-1.0 g wet, ground in liquid nitrogen).
2. Add 750 µl of lysis buffer, stir with a dissecting needle and/or vortex so the mixture is homogenous. Incubate at 65°C, 1 hr.
3. Add 700 µl of chloroform: phenol; 1:1 and vortex briefly. Microcentrifuge at 12,000 rpm for 10 minutes or until the aqueous (top) phase is clear.
4. Remove 600-650 µl aqueous phase to a new tube — be careful not to take any cellular debris from the interface.
5. Add 700 µl of SEVAG, vortex briefly. Be careful, as the caps are loosened by chloroform. Microcentrifuge as above for 5 minutes.
6. Remove the aqueous phase to a new tube (approx. 550-600 µl). Add 20 µl of 3 M NaOAc. Top off the eppendorf tube with isopropanol. Invert gently several times. You should see DNA "ropes" precipitate.
7. Microcentrifuge as above for 30 seconds to pellet the DNA. Pour off the supernatant. Invert the tubes for 1 minute to drain.
8. Add 300 µl TE and place in a heat block at 65° C for 10-15 minutes. Finger vortex to resuspend the pellets.
9. Add 10 µl NaOAc and top off the eppendorf tube with EtOH. Invert gently several times. Nice DNA "ropes" appear again.
10. Microcentrifuge as above 30 seconds to 2 minutes to pellet the DNA. Pour off the supernatant and rinse the pellet with 70 per cent EtOH. Invert to drain 1 minute.
11. Dry the tubes in a vacuum oven at 50°C for 15 minutes at most.
12. Resuspend the DNA pellet in 100 µl of TE and store the tubes at -20°C.

Quantification and Quality Analysis of Genomic DNA

A. Agarose gel electrophoresis method

B. Spectrophotometric method

Dilution of DNA Sample

DNA samples was diluted with distilled water to yield a working concentration of 20 ng/µl and kept at 4°C.

Amplification of 5.8S rDNA gene and ITS regions

Reagents

A. *Preparation of Master mix (50µl/reaction mixture)*

10X PCR buffer	5.0µl
dNTP (25mm)	0.4µl
Primer ITS-1 (50 pmol)	1.0µl
Primer ITS-4 (50 pmol)	1.0µl
$MgCl_2$ (25mM)	2.0 µl
Taq DNA polymerase ($5UmL^{-1}$)	0.2µl
Glycerol (5 per cent)	2.0µl
DH_2O	32.0 µl
Total	44 µl

Note: The 44 µl of master mix in PCR tube and add 6 µl genomic DNA and add few drops of mineral oil before PCR.

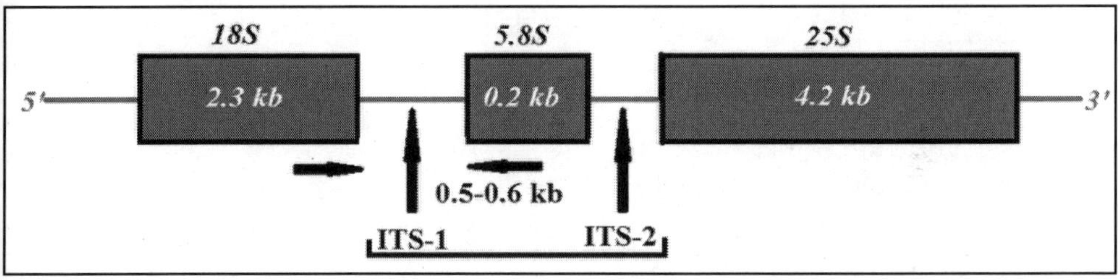

Figure 1: Internal Transcribed Spacer (ITS) Region in Fungal Genome ITS-1= 5'-TCC GTA GAA CCT GCG G-3'ITS-4 = 5'-TCC TCC GCT TAT TAT GC-3'

B. *PCR Conditions: Programme- ITS*

Lid Heating-Enabled

Step1 = 95°C for 1 min (Denaturation step)

Step 2 = 50°C for 30 Sec (Primer Annealing Step)

Step 3 = 72°C for 1 min 20 Sec (Elongation step)

Step-4 = 34 cycle (Repeat Step 1to 3)

Step5 = 72°C for 10 min (Final elongation step)

Hold at 4°C for 1h

C. *Gel Electrophoresis*

The gel run time is 100 min instead of 3 h and voltage is 60 V instead of 45 min.

Restriction Digestion of ITS

In order to develop RFLP profiles of amplified ITS fragments, different restriction enzymes are used to develop very specific RFLP profiles. These profiles serve as a reproducible and reliable method of distinguishing individuals having similar ITS lengths.

Reagents

Buffer (specific to restriction enzyme)	2 µl per tube
Restriction enzyme	1 µl per tube
PCR product of ITS	3 µl per tube
DH_2O	9 µl per tube

Enzymes

Alu 1, Eco R-1, Bsu R-1, Hin f-1, Msp 1

Procedure

1. Incubate after adding the above constituents at 37°C for all the enzymes except Taq, which is heated at 60°C for 1h.
2. Add 50 ul of absolute alcohol to precipitate DNA and incubate tubes at -20°C for 1h.
3. Centrifuge tubes at 18,000 rpm for 10 min at 4°C and discard ethanol (supernatant) without disturbing the pellet.
4. Allow to dry the pellet for 30 min at 50°C and add 10 mL of DH_2O/tube to dissolved the pellet.
5. Use 2 per cent agarose and run the gel at 60 V for 100 min, prepare separate gel for each sample and compare restriction pattern of different sample to catalogue them.

Sequencing of PCR Amplified ITS Product

Sequencing Reaction Dilutions		
Components	*Full Reaction*	*Half Reaction*
Reaction mix (2.5X)	8 µl	4 µl
Template (200ng/µl)	1 µl	1 µl
Dilution buffer (5X)	-	-
Primer (0.8 pmol/µl)	4 µl	2 µl
Millipore water	7 µl	2 µl
Total	20.0 µl	10.0 µl

Big Dye Terminator v 3.1 Clean Up (Tube Method)

The robust, highly flexible chemistry is ideal for *de novo* sequencing, resequencing, and finishing with PCR product, plasmid, and BAC templates.

Sequence Analysis

The resulting ITS sequences were analyzed for homologies to sequences deposited in the GenBank and EMBL databases.

Chapter 4

Microbial Identification by Multi-Locus Sequence Typing (MLST)

(Prassan *et al.*, 2019)

Introduction

The most popular and robust method is rRNA, mostly used for microbial identification and able to differentiate microorganisms below the species level. Multi-locus Sequence Typing (MLST) mostly used to analysis the essential protein coding genes, or housekeeping genes, which encode proteins necessary for the normal cellular functions of the bacterium which represent more variability in their sequence.

Workflow of MLST

This powerful technique used to characterize different isolates of bacterial species through internal fragment sequences from various type of housekeeping genes. The fragment has 450- 500 bp of each gene, can be sequenced on both strands by using Big Dye Terminators v 3.1 chemistries on Applied Electrophoresis systems. For each housekeeping gene, the various sequences present within the bacterial species are specified as distinct alleles. For each isolate, the alleles at all loci define the allelic profile or sequence type (ST).

DNA extraction

↓

16S/ITS2 sequencing and Identification

↓

MLST gene amplification

↓

Sequencing

↓

Phylogenetic analysis

Procedure

The *Burkholderia cepacia* complex MLST scheme uses fragments of the following 7 housekeeping genes.

List of Primers Sequence Used in MLST Analysis of *B. cepacia* Complex

Sl.No.	Gene Name	Primer Sequence
1.	ATP synthase beta chain (*atpD*)	F-GATCGTACAGTGCATCGG R-ATCGTGCCGACCATGTAG
2.	Glutamate synthase large subunit (*gltB*)	F-CGCTCGAAGATCAAGCAG R-GGGAACACCTTCACGAAC
3.	DNA Gyrase subunit B (*gyrB*)	F-CGACAACTCGATCGACGA R-GACAGCAGCTTGTCGTAG
4.	Recombinase A (*recA*)	F-GATAGCAAGAAGGGCTCC R-CTCTTCTTCGTCCATCGCC
5.	GTP binding protein (*lepA*)	F-CGACGGCAAGGTCTACAA R-AGCATGTCGACCTTCACG
6.	Acetoacetyl-CoA reductase (*phaC*)	F-CTCAGCGAATTGCGTACG R-CCGTTCAGCGAGAAGTCG
7.	Tryptophan synthase subunit B (*trpB*)	F-GATCTACCTGAAGCGCGA R-GTGTGCATGTCCTTGTCG

PCR Amplification: PCR Conditions for all primers

Step1 = 95°C for 1 min (Denaturation step)

Step 2 = 50°C for 30 Sec (Primer Annealing Step)

Step 3 = 72°C for 1 min 20 Sec (Elongation step)

Step-4 = 34 cycle (Repeat Step 1to 3)

Step5 = 72°C for 10 min (Final elongation step)

Hold at 4°Cfor 1h

Each 50μl Amplification Reaction Mixture Containing

- ☆ 4.0 µl chromosomal DNA (5-20 ng/µl)
- ☆ 10 µl Q solution (Qiagen)
- ☆ 2.0 µl forward primer (10 pmol/µl)
- ☆ 2.0 µl reverse primer (10 pmol/µl)
- ☆ 0.25 µl Taq polymerase (Qiagen, 5 units/µl)
- ☆ 25. 75 µl PCR grade water
- ☆ 1.0 µl Dntp solution (Qiagen, 10 mM each dNTP)
- ☆ 5.0 µl 10X PCR buffer (Qiagen, 15mM $MgCl_2$)

The amplification product purified by using Min Elute UF plates (Qiagen) following the manufacturers method before being used in a sequencing reaction. Sequencing was started each DNA strand with Big Dye Terminator Ready Reaction Mix v3.1 under standard sequencing conditions. The final reaction products were separated and detected on an ABI PRISM genetic analyser 3100 by using standard sequencing module.

Sequencing Results

Sl.No.	*Gene Name*	*Primer Sequence*
1.	ATP synthase beta chain (*atpD*)	F-GTTCATCTGGCCGTACAC R-AACTGACGCTCGAAGTCC
2.	Glutamate synthase large subunit (*gltB*)	F-CTTCTTCTTCGTCGCCGA R-TTGCCGACGTAGTCGTTG
3.	DNA Gyrase subunit B (*gyrB*)	F-ATCGTGATGACCGAGCTG R-CGTTGTAGCTGTCGTTCC
4.	Recombinase A (*recA*)	F-TGACCGCCGAGAAGAGCAA R-GACCGAGTCGATGACGAT
5.	GTP binding protein (*lepA*)	F-GGCATCAAGGAACTGACG R-CTGCGGCATGTACAGGTT
6.	Acetoacetyl-CoA reductase (*phaC*)	F-AGACGGTATCAAGGTGGT R-ACACGGTGTTGACCGTCA
7.	Tryptophan synthase subunit B (*trpB*)	F-CTGGGTCACGAACATGGA R-CCGAATGCGTCTCGATGA

MLST Data Analysis

A. **Non-redundant databases (NRDB)-** Mostly used to compare batches of sequences to find those that are original. It is useful when developing new MLST schemes and databases does not yet exist.

B. **NRDB align-** This potential tool combined NRDB with EMBOSS show align, taking a FASTA file of aligned sequence as input and displaying the distinct sequences in an interleaved format.

C. **Splits tree-** Used to perform split decomposition analysis on allelic profile data using a web version of Splits tree.

D. **LIAN-Linkage analysis-** This tool used to test the null hypothesis of linkage equilibrium for multilocus data.

E. **Shustring-** This programme mainly for computing Shortest Unique SubSTRINGs.

F. **BURST-** It is clustering algorithm design and used with MLST data.

G. **Tree drawing-** Use the PHYLIP suite of programs to generate neighbor-joining and UPGMA trees from allelic profile data. Tree display and output (in gif, pdf, ps formats) is performed using Phylodendron.

H. **Minimum-spanning tree-** Mainly used Prim's algorithm to draw a minimum-spanning tree from MLST profile data.

Chapter 5

Bioinformatics Approaches in Molecular Data Analysis

(Arjun *et al.*, 2019)

Introduction

Mainly two types of molecular data are generated during molecular identification of bacteria. There are basically two data 16S rRNA based sequence data and DNA fingerprinting data.

A. 16S rRNA Gene Sequencing for Bacteria Identification

Introduction

The potential use of 16S rRNA gene sequence to study bacterial phylogeny and taxonomy has been by far the most common housekeeping genetic marker. Because universal presence in all bacteria, conserved nature of 16S rRNA gene, acting as accurate measure of evolution and sequence length of 16S rRNA gene is used for many bioinformatics based taxonomic annotation pipelines. Various tools can be used to perform 16S rRNA gene sequence based bacterial identification. The tools are listed below:

1. MEGA standalone software for sequence alignment and phylogenetic tree generation
2. Eztaxon web-based tool for identification of prokaryotes
3. Finch TV (chromatogram viewer)

1. MEGA Standalone Software for Sequence Alignment and Phylogenetic Tree Generation

Introduction

Phylogenetic analysis is the study of evolutionary biology of an organism with respect to its related neighbours. The more closely the related species they are more likely to share a recent common ancestor. The primary aim is to discover rates of evolutionary change, predict the sequence function and population history. It helps to identify the changes occurred in the sequences of closely related organisms over a period of their evolution.

Basic steps for construction of phylogenetic tree

1. Choosing sequences
2. Multiple sequence alignment
3. Determination of tree building
4. Assessing the tree reliability

1. Choosing Sequences

For construction a phylogenetic tree can be used either nucleotide or protein sequences. Nucleotide or protein sequences mainly depend on properties of sequences and purpose of study. To appropriate study of evolution of closely related organisms, nucleotide sequence is better and for widely related organisms, either slowly evolving nucleotide sequences like ribosomal DNA and protein sequences can be used.

2. Multiple Sequence Alignment

Critical steps in phylogenetic analysis as it establishes positional correspondence in evolution. Two basic approaches can be used for alignment sequences.

a. **Global alignment:** To analyze evolutionary relationship between distantly related organisms.
b. **Local alignment:** To analyze evolutionary relationship between closely related organisms.

Models

a. Jukes and Cantor Model

This model can be used for multiple mutations per site, which assumes that nucleotides are substituted with equal probability. An appropriate formula for deriving evolutionary distances that include hidden changes is introduced by using a logarithmic function.

$$dAB= -(3/4)\, ln\, [1-(4/3)\, PAB]$$

where,

dAB = Evolutionary distance between sequences A and B

PAB = Sequence distance measured by the proportion of substitutions over the entire length of the alignment

b. Kimura Two-Parameter Model

Most common method used in phylogenetic analysis in which the probability of nucleotide substitution is not equal. A formula used in Kimura model.

$$dAB = -(1/2)\, ln\,(1\text{-pti-ptv}) - (1/4)\, ln\,(1\text{-}2\ \text{ptv})$$

3. Determination of Tree Building Approach

a. **Distance based-** In this approach number of changes between each pair in a group of sequences to construct a phylogenetic tree.
b. **Character based-** Most common approach and uses the sequence characters directly or pair wise distances.

4. Assessing the Tree Reliability Bootstrapping

This is the statistical technique that measures the sampling errors of a phylogenetic tree. A large number of bootstrap resampling steps are needed to achieve a meaningful result and it is generally recommended to have a bootstrap value range from 500 to 1000. The bootstrap values more than 70 per cent which corresponds to a probability of more than 95 per cent that the true phylogeny has been found.

Phylogenetic Tree Construction by Using MEGA (6.0) Software

MEGA (Molecular Evolutionary Genetic Analysis) software is most valuable method used for tree construction, it is user friendly and easy to operate for construction of phylogenetic tree. The nucleotide sequences of very closely related strains of species are retrieved from the database. By using MEGA 6.0 software, phylogenetic tree is constructed of close related species by maximum parsimony algorithms, maximum-likelihood and neighbour-joining.

a. Aligning Sequences by using Clustal W

This creates multiple sequence alignment in MEGA through the Clustal W and Muscle.

Procedure

1. Open file in FASTA format and select alignment/align by Clustal W from the main menu to align the selected sequences data.
2. Click the OK button to accept the default settings for Clustal W.
3. After alignment is fully complete and then save the current alignment session by selecting data/export alignment/MEGA format from the main menu.
4. Give the file name such as 'Sequence'. This will allow the current alignment session to be restored for future editing and tree construction.
5. Exit the alignment explorer by choosing data/exit aln explorer from the main menu.

b. Tree Construction by Neighbour-Joining Method

Procedure

1. Open file and select the saved file (MEGA format) from the appropriate folder.
2. Click phylogeny and select construct/test Neighbour Joining tree option.
3. Test of phylogeny bootstrap method, No. of Bootstrap replications range from 500- 1000.
4. Substitutions type nucleotide, model Kimura two-parameter model, substitutions to include transitions+ transversions, Gaps/missing data treatment partial deletion, site coverage cut-off 95 per cent.
5. Click compute for construction of phylogenetic tree.

Chapter 6

Total Protein Extraction with TCA-Acetone and 2D Gel Electrophoresis

Materials and Equipment Required

- ☆ Electrophoresis apparatus
- ☆ Electrophoretic tank
- ☆ Casting tray
- ☆ Comb
- ☆ Micropipettes
- ☆ Conical flasks for gel preparation
- ☆ Spectrophotometer
- ☆ Spatula

Reagents Required

- ☆ Precipitation solution: 10 per cent TCA (w:v), 0.07 per cent 2ME (v:v) in cold acetone. This solution must be prepared freshly and stored at - 20°C until use. All three components are toxic and the solution must be prepared under a hood.
- ☆ Rinsing solution: 0.07 per cent 2ME (v:v) in cold acetone. This solution is then stored at - 20°C for a month.
- ☆ R2D2 solubilization solution: Urea (5 M), thiourea (2 M), CHAPS 2 per cent (w:v), SB3-10 (2 per cent) (w:v), DTT (20 mM), phosphine (5 mM), pharmalyte (0.5 per cent) 4–6.5 (v:v), pharmalyte 3–10 (0.25 per cent) (v:v) in ddH_2O. The solution is heated (below 30°) to help urea solubilize. It is aliquoted and then stored at - 80°C for months.

- ☆ UKS solubilization solution: urea (9.5 M), K_2CO_3 (5 mM), SDS (1.25 per cent), DTT (5 per cent), Triton X-100 (6 per cent), ampholines (2 per cent) 3.5-9.5 in ddH_2O. K_2CO_3 is prepared as a 2.8 per cent (w:v) stock solution, SDS as a 10 per cent (w:v) filtered stock solution, and Triton X-100 is provided as a 20 per cent solution. Besides 3 mL ddH_2O are added to other components till the urea is solubilized (heating below 30 °C). A 40 mL solution is usually prepared, then aliquoted and stored at -80°C for months.
- ☆ IPG strip rehydration solution: Solution A: 7 M urea, 2 M thiourea, 1.4 per cent CHAPS (w:v), 16 mM DTT, 5 mM phosphine, 0.3 per cent pharmalyte 3-10 (v:v) in ddH_2O. Solution B: 7 M urea, 2 M thiourea, 0.5 per cent CHAPS (w:v), 10 mM DTT, 5 mM phosphine in ddH_2O.

Procedure

1. *Protein Precipitation and Denaturation*

- ☆ Plant tissues are ground in mortar with a pestle in liquid nitrogen and a fine powder is thus obtained.

  Note: A very fine powder should be obtained; it may be obtained by the pre-crushing of hard materials. An automatic cryogenic crusher with a metallic ball may be used to conduct this process.
- ☆ About 200 µL of powder is transferred to a weighted 2-mL eppendorf tube, and covered with 1.8 mL of the cold TCA-2ME-acetone solution, mixed, and then stored at - 20°C for 1 h. TCA and acetone have the capacity to denature and precipitate the proteins. This solution inactivates the phenoloxidases and oxidases, preventing phenol oxidation into quinones, which would result in protein binding to insoluble complexes. It was also shown to inactivate proteases, as well as phenol extraction. Acetone alone allows the solubilization of the pigments, lipids, and terpenoids possibly present in the tissue. 2ME prevents the formation of disulphide bonds during precipitation.

  Note: It is important to work the procedure at a low temperature to limit protease action.
- ☆ Centrifuge for 10 min at 10,000 rpm (refrigerated centrifuge, below 4°C).

2. *Rinsing with 2ME-acetone Solution*

- ☆ The supernatant is then discarded and the pellet is resuspended in 1.8 mL of cold rinsing solution. It should be stored at - 20°C for 1 h. This step is critical in the sense that it can eliminate the acidity due to TCA that would impede protein recovery.
- ☆ Centrifuge 10,000 rpm for 15 min at a low temperature of 4°C.
- ☆ The supernatant is then discarded and the step is then repeated twice.
- ☆ The pellet is dried under vacuum for about 1 h to eliminate acetone.
- ☆ Weigh the pellet.

3. Protein Solubilization

- ☆ The amount of UKS or R2D2 buffers for protein solubilization depends on the plant tissue; for instance, we use 60-µL/mg dry powder for leaf (maize, rape) and 50-µL/mg dry powder for maize kernel. Re-solubilization is achieved by vertexing for 1 min.
- ☆ Centrifuge at 10,000 rpm (at 25°C) for 15 min and collect the supernatant in a 1.5 mL eppendorf tube.
- ☆ Centrifuge again (15 min, 25°C), and the supernatant is then transferred to a new eppendorf tube. Samples containing solubilized proteins can then be stored at - 50°C or - 80°C for months.

4. Preparation of Samples for IEF

- ☆ Before using the protein samples, they must be centrifuged again. The pellet is then discarded.
- ☆ Around 50 µg of total proteins are loaded per IEF gel (24 cm long, either classical rod gels or IPGs) for analytical gels, which are then revealed with silver nitrate, or 150–500 µg for gels which are revealed with colloidal Coomassie blue.
- ☆ Samples solubilized in R2D2 are suitable for use in IEF with IPGs. The required volume of solubilized samples should be complemented to reach the desired protein amount with the R2D2 solubilization buffer, up to 450 µL for active rehydration.

Safety Considerations and Standards

- ☆ The precipitation and rinsing solutions must be cold when used. It is important to keep an acetone bottle at 4°C temperature to prepare these solutions.
- ☆ Rehydration buffer contains high molarity of urea, so it must not be heated above 30°C as isocyanate ions are produced, which result in protein carbamylation.
- ☆ For efficient protein extraction, it is important to obtain it in fine powder. It can be achieved by the pre-crushing of hard material (*e.g.*, mature maize grains). An automatic cryogenic crusher with a metallic ball (6 mm diameter) can also be used for crushing.
- ☆ To recover a white pellet, various rinsing steps can be applied with highly pigmented samples. It is feasible to extend rinsing overnight.
- ☆ The dry pellet powder can also be stored at - 80°C before protein solubilization. However, in this case, it is better to re-dry the powder prior to protein re-solubilization.
- ☆ It is important to handle all the reagents involved with gentle care.

Chapter 7

Gene Cloning/ Recombinant DNA Technology (Masoodi *et al.*, 2020)

Introduction

The process of making multiple exact/identical copies of a specific gene or piece of DNA (clones) using the techniques of genetic engineering is known as gene cloning/DNA cloning/molecular cloning/recombinant DNA technology. This method is particularly used for gene isolation and its amplification.

Materials and Equipments

Host- in which the gene of interest is inserted/recombinant DNA replicates and it should be viable, *e.g.*, E*scherichia coli*-DH5α

Vector- to carry, maintain, and replicate the gene of interest in the host cell, *e.g.*, bacterium, plasmids, bacteriophages, bacterial artificial chromosomes (BACs), mammalian artificial chromosomes (MACs), *etc.*, *e.g.*, pEGFPC1.

Gene of Interest- the desired genes to be cloned.

Attributes of an Ideal Vector

An ideal vector should possess the following:

Marker for selection- which helps to distinguish the vector containing cells from other cells, *e.g.*, Antibiotic resistance gene.

Origin of replication- a point at which DNA replication starts, which facilitates DNA opening, *i.e.*, self-replicating nature. This site is usually AT rich and contains restriction and ligase enzymes.

Small size. *e.g.*, 4.7 kb.

Multiple/poly cloning site (MCS/PCS)- a DNA region within a plasmid containing multiple, specific, distinctive restriction enzymes cut sites or recognition sequences, where a number of restriction enzymes can cut or cleave after recognizing their recognition sites.

High copy number- to produce more copies of the gene of interest, *e.g.*, recombinant strain of *E. coli* DH5α contains a high copy number of plasmids due to the presence of a mutated RoP gene.

Marker for recombination- which helps in Blue-White screening/selection of clones, *e.g.*, lac Z gene in *E. coli.*

Introduction of donor DNA fragment must not interfere with the replication property of the vector.

- ✰ Its isolation from the host cell should be easy.

Attributes of a Good Host Cell

- ✰ It should permit the comfortable entry of the recombinant DNA into the cell.
- ✰ Should not possess endogenous DNA deteriorating enzymes and shall not destroy the recombinant DNA it uptakes.
- ✰ Should be able to stably maintain and contain the recombinant DNA.
- ✰ The transformed host must not independently sustain outside the laboratory.
- ✰ Should be easy to maintain and handle.
- ✰ Should be available in a wide range of genetically defined strains.
- ✰ Should accept a large cohort of different vectors.

Procedure

- ✰ Isolation of Vector from *E. coli.*
- ✰ Isolation of gene of interest to be cloned. DNA in case of prokaryotes and mRNA in case of Eukaryotes.
- ✰ Primer designing for amplification of gene of interest. The primers should have restricting sites for enzymes that are in MCS of vector. Selection of Restriction sites on the primer should be such that these restriction sites should NOT be anywhere in the gene part to be amplified.
- ✰ If cloning is for expression, primers shall be such that the start signal of the gene sequence should be in frame with the vector.
- ✰ PCR amplification of gene of interest.
- ✰ Gel elution of gene of Interest after PCR amplification.
- ✰ Restriction digestion of vector and gene of interest. A second gel elution of restricted gene of interest and restricted vector.
- ✰ Ligation of vector and insert.

- ☆ Transformation of recombinant DNA molecule into a suitable host cell.
- ☆ Selection of host cells to be transformed and identification of the clone which contains the gene of interest.
- ☆ Multiplication/expression of the introduced gene in the host.
- ☆ Isolation of multiple gene copies or protein expressed by the gene.
- ☆ Purification of the isolated gene copy or protein.

In the detailed example below, we have described the cloning of a CBF gene from Tomato in the pEGFPC1 vector to make a fusion protein of GFP-CBF1.

Plant Material

Tomato seeds were collected from seed repositories of SKUAST-Kashmir. The seed material was soaked in water for 10 min, washed twice with sterile water, and then sterilized seeds were germinated in Petri plates at 24°C in the incubator. Germinated seedlings were transferred to Petri plates and grown under controlled conditions in a growth chamber at 24°C/15°C (Day/Night) and were processed for RNA extraction.

Microbial Cultures and Plasmids

E. coli (DH5α) was obtained from the National Institute of Immunology, New Delhi, India. The pEGFPC1 was kindly supplied by Prof Zhou Wang, University of Pittsburgh, Pennsylvania, USA.

DEPC Treatment

Microfuge tubes, tips, tip boxes, mortar and pestle, and spatula were DEPC treated overnight and then autoclaved followed by drying in a hot air oven.

RNA Extraction and qRT-PCR

Total RNA was extracted at the germination stage from the seedlings of tomato using the Trizol (Invitrogen) (Chomczynski and Sacchi, 1987) method as follows.

Tissue samples (100 mg) were ground in a mortar and pestle in liquid nitrogen and transferred to a new microfuge tube. To it, 1 mL of Trizol was added to the homogenized sample and mixed thoroughly and incubated at room temperature for 5 min; 0.2 mL of chloroform was added and the tubes were shaken vigorously for 15 s. The tubes were then incubated at room temperature for 2–3 min. The samples were centrifuged at 12,000 rpm for 15 min at 4°C. The mixture was then separated into lower red, an interphase, and a colorless upper aqueous phase. The aqueous phase exclusively contains RNA. The upper aqueous phase was carefully transferred into a new 1.5 mL tube without disturbing the interphase. The aqueous phase was processed for RNA precipitation. To this aqueous phase 0.5 mL of 100 per cent isopropanol was added. Incubation at room temperature for 10 min was done. Centrifugation was done at 12,000 rpm for 10 min. The supernatant was removed from the tube with the RNA pellet remaining inside. The pellet was then washed with 1 mL of 75 per cent ethanol. The samples were vortexed briefly and then the tubes were centrifuged at 7500 rpm for 5 min at 4°C. The RNA pellet was then air-dried for 5–10 min and it was resuspended in RNase free water (DEPC water). After dissolving, samples were run on a 1 per cent agarose gel to analyze the quality of RNA and the presence of any DNA contamination.

DNase Treatment of RNA Samples

The Thermo Scientific DNase kit was used for removing traces of DNA in the extracted RNA. The DNase treatment was given following the manufacturer's protocol.

Quantitative and Qualitative Analysis of RNA

The quality of RNA was determined by using Nanodrop (*Thermo Fisher SCIENTIFIC*) at an OD of 260 and 280 nm. The samples showing an OD260/280 ratio between 1.9 and 2 were used for further experimental studies. RNA quantity was estimated by using an OD of 260 nm.

cDNA synthesis

cDNA synthesis was done using the Thermo Scientific Revert Aid First Strand cDNA Synthesis Kit using oligo dT primers. The first strand cDNA was synthesized by using the manufacturer's protocol and then stored at - 20°C. The protocol for the synthesis of first strand is given below.

The following reagents were added in a tube free from the RNase free tube on ice as per the given order.

Template	Total RNA	1.5 µg
Primers	Oligo $(dT)_{18}$ primer	1 µL
Nuclease free water		Final vol to 12 µL
Total volume	**15 µL**	

The components were gently mixed and a brief spin was given. Incubation at 65°C for 5 min was given. The tube was chilled on ice and a brief spin was given. The tube was again placed on ice and the following components were added as per the indicated order.

5 × reaction buffer	4 µL
RiboLock RNase inhibitor (20 U/µL)	1 µL
10-mM dNTP mix	2 µL
RevertAid M-MuLV RT (200 U/µL)	1 µL

The components were gently mixed and a brief spin was given. The components were incubated at 42°C for 60 min. The reaction was terminated by heating at 70°C for 5 min.

Primer Designing

For designing primers for amplification of genes coding for CBF1 from *Solanum lycopersicon*, amino acid sequences for the previously submitted GenBank sequence were retrieved from GenBank (Farheena, 2018—M.Sc. Thesis). The mRNA sequence was aligned using Clustal W (www.ebi.ac.uk/clustalw) to check for conserved motifs. Two oligonucleotide primers for the CBF1 gene were designed using the primer 3 online software. A restriction site was added in the primers at the 52 and the 32 ends.

Validation of cDNA and Primers using PCR

Polymerase chain reaction using primers of the GAPDH gene was carried out to validate the cDNA synthesis and primers. Details of the primers used are given in the following table. Primers for amplification of CBF-1 and GAPDH genes. The bold underlined sequence represents the site for the restriction enzyme.

Nature of Primer	*Primer Sequence*		*Size of Amplicon*	*Tm (°C)*	*Restriction Site*
GAPDH	Forward primer	52 CCTTAGTTG ATTCTGTACC32	169 bp	59.7	Nil
	Reverse primer	52 GCCAGTTAGT AGTAATTGCCA 32		58.3	Nil
CBF1	K-Lab-CBF1-LS-F	CTCAGTAGATCTATG AATATCTTTGAAACCTA	618 bp	55	*Bgl*II
	K-Lab-CBF1-LS-R	CGTACTCTGCAGTTA GATAGAATAATTCCATA		55	*Pst*I

PCR Reaction Mixture

PCR amplification for CBF-1 and GAPDH genes was carried out in a reaction volume of 20 µL in 0.2 mL PCR tubes. The reaction mixture contents are given in the table below.

PCR Reaction Mixture for Amplification of CBF-1 and GAPDH Genes

Constituents	*Volume*
Autoclaved water	13.6 µL
PCR buffer (1 ×)	2 µL
$MgCl_2$ (25 mM)	1.2 µL
dNTPs (25 mM)	0.4 µL
Forward primer (10 µM)	1 µL
Reverse primer (10 µM)	1 µL
Taq Polymerase (5 U/µL)	0.3 µL
cDNA sample (70 ng/µL)	0.5 µL
Total volume	20 µL

PCR Cyclic Conditions

The amplification reaction was carried out in a thermocycler (*Applied Biosystems*). The cycling conditions are shown in the following table. Thermal cycling conditions for amplification of CBF-1 and GAPDH genes.

Steps		*Temperature (°C)*	*Time*	*Number of Cycles*
1. Hot-Start		94	2 min	1
2. Denaturation		98	10 s	35
3.Annealing	GAPDH	61	30 s	
4. Extension		68	2 min	
5. Final extension		68	10 min	1

Gel Electrophoresis

The amplified products of GAPDH and CBF1 genes were electrophoresed on a 2 per cent agarose gel and compared with the 100 bp DNA ladder respectively.

cDNA Quantification

cDNA quantification was done using a Qubit® 2.0 Fluorometer (Invitrogen).

Confirmation of cDNA Synthesis

The internal gene primers for GAPDH used for confirmation of cDNA synthesis were designed (Table 1) using the protocol outlined in the primer3 online software. The primers were checked for folds and self-complementarity by oligocalc software. These GAPDH primers were used for confirmation of cDNA synthesis. The predicted size of the amplicon was 169 bp.

Primers for Amplification of CBF1 and GAPDH Genes

Nature of Primer	*Primer Sequence*		*Size of Amplicon*	*Tm (°C)*
GAPDH	Forward primer	52 CCTTAGTTG ATTCTGTACC 32	207bp	59.7
	Reverse primer	52 GCCAGTTAGTA GTAATTGCCA 32		58.3

Gene Cloning

CBF1 was cloned from cDNA of *Solanum lycopersicon* Mill.

PCR Amplification of Full-length Genes (CBF-1)

Total RNA was extracted from 1- to 2-week-old seedlings of *Solanum lycopersicum* Mill. var. Pusa Sheetal. First strand cDNA was prepared with 2 μg of total RNA.

RNA isolation and cDNA synthesis from *Solanum lycopersicum* (Pusa Sheetal)

⇩

PCR amplification of *CBF1* from *cDNA*

⇩

Restriction digestion of *CBF1* and pEGFP-C1 vector with BglII/PsTI

⇩

Gel elution of the digested fragments

⇩

Ligation of *CBF1* to pEGFP-C1 vector with T4 DNA ligase

⇩

Transformation of the ligation mixture into bacterial competent cells (DH5a)

⇩

Screening of the transformants for the *CBF1*-bearing pEGFP-C1 by plasmid isolation and restriction digestion

Strategy from Cloning of CBF-1 in pEGFPC-1.

The putative full-length *CBF1*cDNAs were amplified with K-Lab-CBF1-LS-F and K-Lab-CBF1-LS-R primers. After an initial denaturation at 94°C for 2 min, 35 cycles were run each with 30 s of denaturation at 94°C, followed by 40 s annealing at 55°C and a 1 min extension at 72°C.

The following constituents were mixed gently and the tubes were spin briefly:

Constituents	*CBF1*
Buffer (10 ×)	2.5 µL
dNTP's (2.5 mM each)	2.5 µL
$MgCl_2$ (25 mM)	1.0 µL
cDNA (PB)	1.0 µL
Vent DNA Pol.	1 U
Gene specific primers F + R (10 mM)	2 + 2 µL

The PCR reaction was run in an eppendorf thermocycler as per the following protocol.

The protocol followed for the PCR is as follows:

Sl.No.	*Gene*	*CBF1*	
	Step	*Temperature (°C)*	*Time (min)*
1	Initial denaturation	95	3:00
2	Denaturation	95	3:00
3	Tm	50	0:30
4	Elongation	72	1:30
5	Step 2 repeat 30 cycles		
6	Final extension	72	10:00
7	Hold	4	∞

The products were run on a 1 per cent agarose TAE gel and visualized on a UV transilluminator. Photographs were taken using a gel documentation system. A 1 kb/100 bp ladder (Genetix) was used as a molecular weight marker. The DNA bands were excised from the gel with a sharp sterilized blade, weighed, and put in an autoclaved 2 mL microfuge tube. DNA purification from the excised band was carried out by using the MinElute gel purification kit (QIAGEN) according to the manufacturer's instructions.

Restriction Digestion of the PCR Fragments and Vector

The eluted fragments of CBF1 and pEGFPC1 vector were double digested simultaneously with the specific restriction enzymes using the following protocol.

Constituents	*CBF1 (µL)*	*pEGFPC1*
DNA	5	5µL
Buffer (5×)	2	2µL
Restriction enzyme (thermo) (1 U)	BglII and *Pst*I	BglII and *Pst*I
H_2O	1	1µL

The above constituents were gently mixed and the tubes were spun briefly and incubated at 37°C for 1 h. The products were run on a 1 per cent agarose TAE gel and visualized on a UV-transilluminator and photographed using a gel documentation system. A 1 kb ladder was used as a molecular weight marker. The restricted fragments were gel purified using the MinElute gel purification kit (QIAGEN) according to the manufacturer's instructions.

Expression Vector

The expression vector used was pEGFPC1. Vector preparation was carried out by double digestion of pEGFPC1 expression vectors using restriction enzymes as given in the Table above.

Ligation

The ligation reaction was performed using 0.2 µg of the digested vector, 0.4 µg insert, 1 µL ligation buffer, and 1 U T4 DNA ligase in a final volume of 10 µL. The samples were incubated at 4°C for 24 h and the recombined plasmid was transformed into *E. coli* DH5Q and BL21 (DE3).

Transformation

A single colony of *E. coli* DH5Q was inoculated into 10 mL LB broth in a 50 mL falcon tube. It was grown at 37°C for 16 h. 1 mL of the above grown culture was used to inoculate 100 mL of LB in a 250 mL flask and kept in an incubator shaker at 37°C for 4 h with constant shaking (250 rpm). The culture was kept on ice for 10 min and transferred to a 50 mL falcon tube and centrifuged at 5000 rpm for 5 min at 4°C. After centrifugation, the supernatant was decanted and the cells were resuspended in 1 mL cold 0.1 M $CaCl_2$. The cells were vortexed and again centrifuged at 5000 rpm for 5 min at 4°C. Centrifugation was conducted and the supernatant was decanted. The cells were resuspended in 1 mL cold 0.1 M $CaCl_2$. The cells were vortexed and incubated on ice for 20 min to make them competent. The competent cells were dispensed in 2 mL microfuge tubes (200 µL/tube) and stored at - 80°C for further use.

5 µL of the ligation mixture was pipetted gently into 200 µL of competent cells. The tubes were incubated for 30 min on ice, transferred for 2 min to 42°C (heat shock), and immediately put back on ice. 900 µL of LB was added and the tubes were incubated at 37°C for 2 h. 100 µL of the cells were plated on LB Ampicillin (50 µg/mL)/plates. 100 µL of the control DNA was also plated on LB Ampicillin (50 µg/mL) plates. The plates were grown over night at 37°C in an incubator.

Confirmation of Cloning by Restriction Digestion

The clones were separately inoculated in 5 mL LB broth and incubated for 16 h with 250 rpm constant shaking at 37°C. Plasmid DNA was isolated from clones using the plasmid purification kit (Sigma) following the manufacturer's instructions. The plasmids were stored at - 20°C for further use. Restriction digestion of the vectors was carried out by the following protocol:

Gene	*CBF1*	*Empty Vector*
Plasmid clone (0.5 µg)	pKZ-CBF1	pEGFPC1
Buffer (1.5 µL)	Tango	Tango
R. E. - 1	BglII	BglII
R. E. - 2	PstI	PstI

The samples were incubated at 37°C for 4 h for complete digestion. Enzymes were inactivated at 65°C for 10 min. The products were run on a 0.7 per cent agarose TAE gel at 8 V/cm for 30-45 min. The products were visualized and photographed.

Heterologous Expression

The expression vector pEGFP-C1 (Clontech) was used to generate fusion protein constructs with GFP at the N terminus of CBF1 for convenient visualization using fluorescent microscopy.

Cell Culture Experiments

Human C4-2 prostate cancer cells were obtained from Dr. Zhou Wang, University of Pittsburgh, PA, USA and maintained in RPMI1640 medium supplemented with 10 per cent FBS, 1 per cent glutamine, 100 units/mL penicillin, and 100 mg/mL streptomycin (Invitrogen) at 37°C in the presence of 5 per cent CO_2 in a humidified incubator. GFPCBF1 and GFP expression vector was transiently transfected into C4-2 cells using Lipofectamine Reagent according to the manufacturer's protocol (Thermo). Cells were transfected at > 60 per cent confluence in phenol red-free OptiMEM. The Expression of GFP-CBF1 fusion proteins was imaged 16 h after transfection using the LMI Fluorescent Microscope, ABE, London, UK.

Expected Outcomes

- ✰ Multiple identical copies of gene of interest.
- ✰ Functional GFP fusion protein.

Advantages

- ✰ Facilitates isolation of a particular gene and determination of its nucleotide sequence.
- ✰ Helps in identification of control sequences of DNA.
- ✰ Helps in analyzing the control sequences of DNA.
- ✰ Helps in investigating the functions of RNA/proteins/enzymes.
- ✰ Facilitates the identification of mutations. *e.g.*, gene defects
- ✰ Facilitates the engineering of organisms for specific purposes. *e.g.*, insect resistance, insulin production, *etc.*

Limitations

- ✰ Comes with a degree of uncertainty.
- ✰ Has the tendency to bring new diseases.
- ✰ Might lead to problems in organ rejection.
- ✰ Gene diversity may be decreased.

Chapter 8

Scanning Electron Microscope (SEM)

Introduction

Scanning Electron Microscope (SEM) is a type of electron microscope that scans surfaces of microorganisms that uses a beam of electrons moving at low energy to focus and scan specimens. SEM has been used as an important tool for giving a three dimensional image of the surface of the specimen and is for viewing large objects. It is mostly useful for better understanding the plant microbe interaction in perspective of biological control. SEM can give advantages namely high resolution and large depth of field. It operates under high voltage electron beam, hence surpasses light

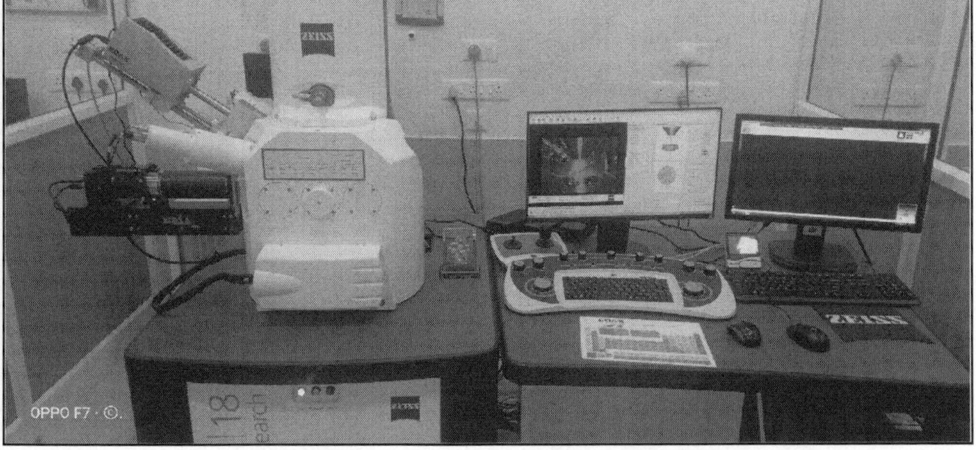

Figure 8.1: Picture of Scanning Electron Microscope Working Principles.

microscope in which higher magnification cannot be achieved due to the limitations of photons/light. The magnification ranges of SEM from × 10 to × 200,000 which allows spectacular analysis of the particular sample. To achieve higher magnification the whole microscope column, including specimen chamber, operates under high vacuum to prevent gas to interrupt either the incident or scattered electron.

The Scanning electron microscope works on the principle of applying kinetic energy to produce signals on the interaction of the electrons. These electrons are secondary electrons, backscattered electrons and diffracted backscattered electrons which are used to view crystallized elements and photons. Secondary and backscattered electrons are used to produce an image. The secondary electrons are emitted from the specimen play the primary role of detecting the morphology and topography of the specimen while the backscattered electrons show contrast in the composition of the elements of the specimen.

How Scanning Electron Microscopy Works

At the heart of a scanning electron microscope is a high-energy electron source positioned above a series of condenser lenses and apertures which focus these electrons into a beam. The position of this beam is altered by sets of deflection or scanning coils before the final lens aperture. A sample is placed in the path of the electron beam which is continuously deflected into a raster scanning pattern by the deflection coils.

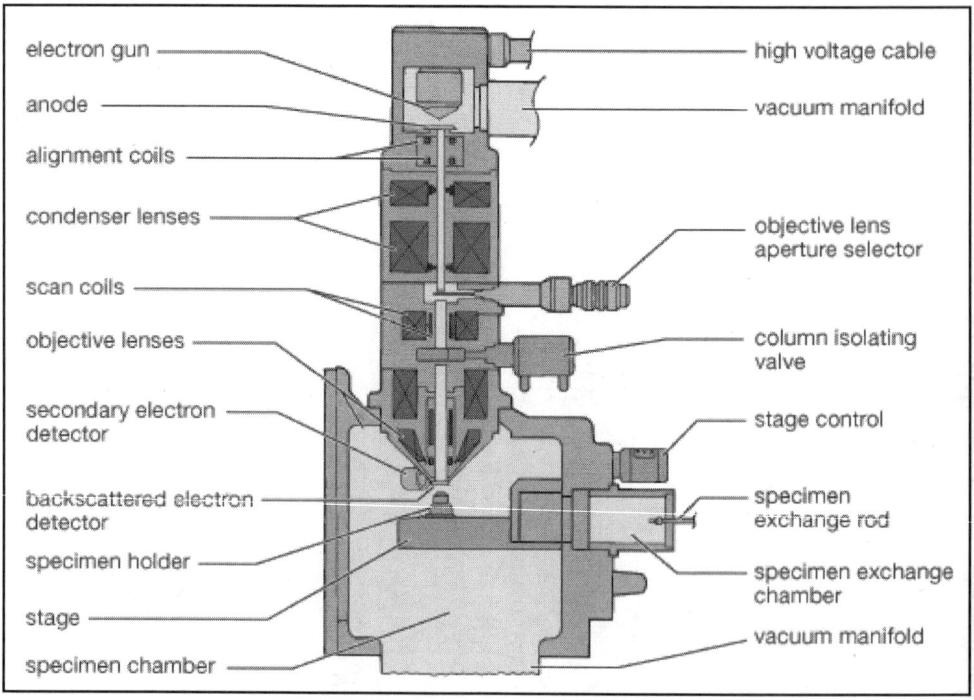

Figure 8.2: Various Parts of Scanning Electron Microscope.

When electrons impact a surface, they generate secondary and backscattered electrons (BSE), as well as x-rays. BSE and x-ray detectors in the sample chamber acquire these signals, which are characteristic of the sample's elemental composition, morphology, and crystalline structure. Scanning electron microscopy can subsequently be used for imaging the elemental composition of a sample surface and determining topographical sample features with a significantly increased resolving power.

Scanning electron microscopy can generate 3D chemical surface maps of a sample with a magnifying capacity of up to 50,000x. This offers high lateral resolution ranging from millimeters to nanometres (>10nm), while energy dispersive X-ray (EDX) analysis provides chemical detection limits of 1000 – 3000 parts per million (ppm).

Sample Preparation for Scanning Electron Microscope (Biological Specimen)

1. Cleaning the Surface of the Specimen

The proper cleaning of the surface of the sample is important because the surface can contain a variety of unwanted deposits, like dust, media components, and other contaminants, depending on the source of the biological material.

2. Stabilization/Fixation

Currently there are various ways of stabilizing a biological specimen. Stabilization is typically done with fixatives, for instance perfusion and microinjection, immersions, or with vapors using different kinds of fixatives *i.e.*, aldehydes, osmium tetroxide, tannic acid and thiocarbohydrazide. Chemical fixation can be done by immersing the specimen in a 1.5 per cent glutaraldehyde solution prepared in 0.1 M cacodylic acid buffer (pH 7.3) and incubated at 4°C overnight. Post fixation can be achieved by osmium tetroxide, improve bulk conductivity of the specimen, but not necessary provide a better fixation.

3. Rinsing the Specimen

After completion of fixation, specimens must be rinsed in order to remove the excess fixative. Samples rinsed in 0.1 M cacodylic acid buffer (pH 7.3), done with one time for 10 min, and then three times for 20 min at 4°C.

4. Dehydration of Specimen

The dehydration process of biological sample needs well skill person and very carefully process. A fixed specimen is dehydrated by incubation in a graded series of ethanol or acetone solutions. Immersion of the specimen can be done by series of acetone that is, 50 per cent Acetone for 5 min, 70 per cent acetone for 10 min, 80 per cent acetone for 10 min, 90 per cent acetone for 15 min, and 100 per cent acetone at 4°C. The solvent concentration is increased gradually so that water is removed gently, without causing specimen shrinkage.

5. Drying the Specimen

SEM operates under vacuum. Sample must be dry or the sample will be destroyed in the electron microscope chamber. According to various microscopists consider a process called the Critical Point Drying (CPD) as the gold standard for SEM sample drying. Carbon dioxide (CO_2) is removed after its transition from the liquid to the gas phase at the critical point, and the sample is dried without any structural damage.

6. Mounting the Specimen

After completion of such process that is, cleaned, fixed, rinsed, dehydrated, and dried using an appropriate protocol, samples must be mounted on a holder that can be inserted into the

SEM. Samples are mostly mounted on metallic (aluminium) stubs with the help of double sticky tape. Moreover, silver-containing glue can additionally be applied for even more conductivity.

7. Sputter Coating with Conductive Material

The reason behind the coating the specimen is to enhanced its conductivity in the SEM and to prevent the build-up of high voltage charges on the sample by conducting the charge to ground. The metal is applied in a controlled manner in a sputter coater. Mostly, samples are coated with the help of conductive heavy metal material that is, gold, platinum, gold- palladium, and these materials have a very thin layer ranges from 20 nm to 30 nm. In case, coating is thick enough to prevent charging but not thick enough to obscure specimen surface details.

Flowchart of Protocol for Specimen Processing for SEM

1. Primary fixation (2.5 per cent Glutaraldehyde)

Take an Eppendorf tube (2 mL), add 400 µl of 2.5 per cent Glutaraldehyde and then add2-4 loopful of biological sample.

Incubate at room temperature 1 h and then kept for overnight at 4°C.

↓

2. Washing (0.1 M phosphate buffer)

Take overnight incubated samples, centrifuge at 5000 rpm for 5 min. Discard the supernatants and add 400 µl of phosphate buffer.

Mix well and centrifuge at 5000 rpm for 5 min. Repeat this step for two times.

↓

3. Post fixation (optional, 1 per cent OsO_4/1 per cent $AgNO_3$)

Take the pellet and add 400 µl of freshly prepared 0.1 per cent $AgNO_3$ and mix it in the sample.

Incubate at room temperature for 1 h.

4. Washing (0.1 M phosphate buffer)

Centrifuge at 5000 rpm for 5 min.

Discard the supernatants and add 400µl of distilled water. Repeat this step for two times to remove excess $AgNO_3$.

5. Dehydration (All steps conducted at 4°C)

Take pellet and add 400µl of 30 per cent alcohol and incubate at 4°C for 15 min. Centrifuge at 5000 rpm for 5 min, discard the supernatant and take pellet.

Repeat this step.

Finally take pellet and add 400 µl of 100 per cent alcohol and incubate at 4°C for 1 h and keep the sample in 100 per cent alcohol until further used.

6. Critical point drying (CPD)

It is done with the help of liquid carbon dioxide at its critical point, *i.e.*, 31.5°C at 1100 p.s.i.

7. Mounting

Mount the specimen on to the aluminium stubs.

8. Coating

Used silver or gold material, coat the samples with 20-30 nm thick film with the help of sputter coater machine.

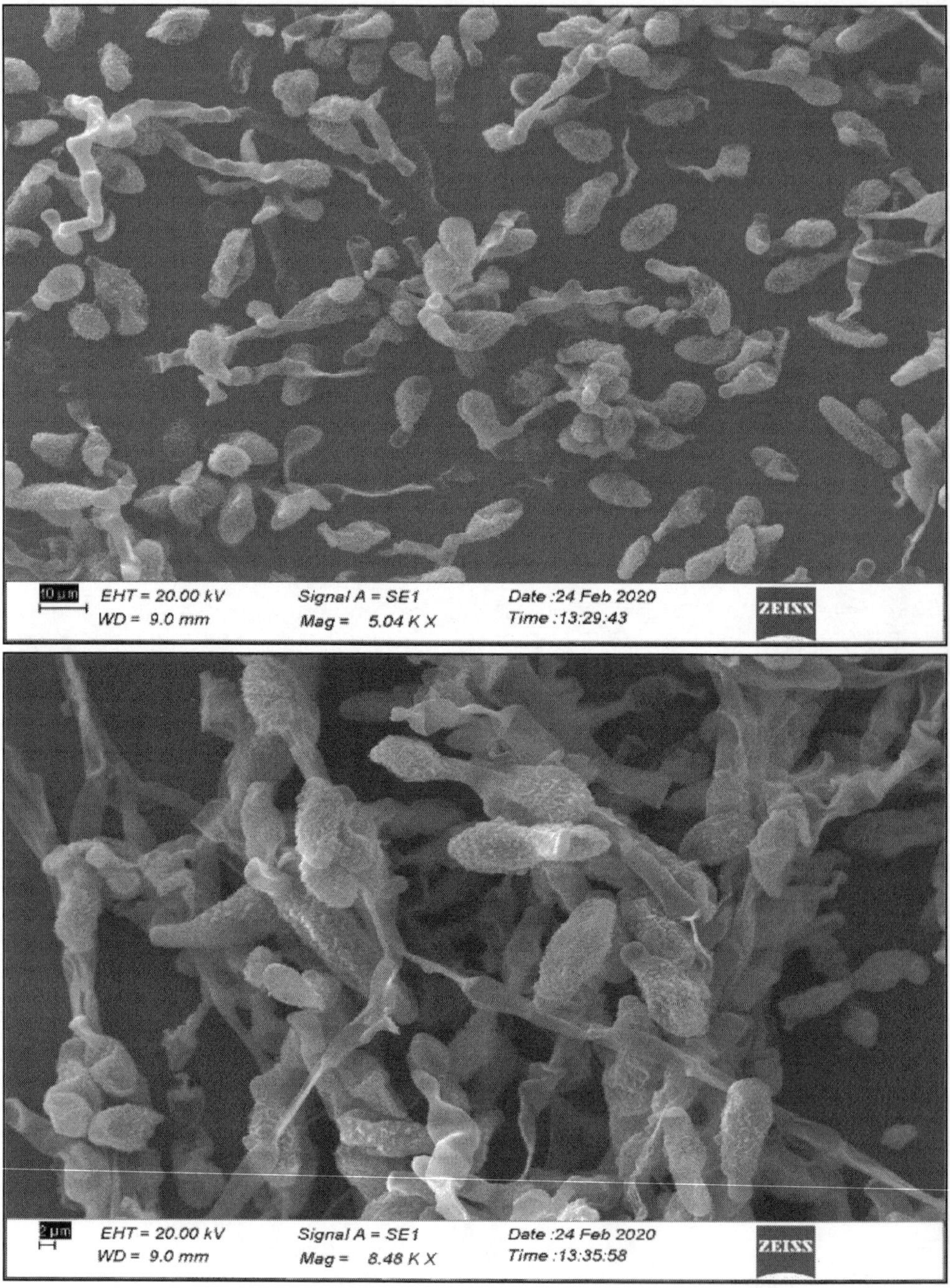

Figure 8.3: Electron Micrograph Showing *Alternaria solani* Fungus Spores.

Chapter 9

Transmission Electron Microscope (TEM)

(Lee *et al.*, 2012)

Introduction

Transmission electron microscopy is a potential technique in which a beam of electrons is transmitted through a sample and used for observing the very fine details of organelles in cells or tissues. This protocol is to be used to exam the membrane structure in cells with or without pathogen infection. The first TEM was demonstrated by Max Knoll and Ernst Ruska in 1931. In 1986, Ruska was awarded the Nobel Prize in physics for the inventing the TEM.

How do TEM Work

Transmission electron microscope used high voltage electron beam to produce high quality pictures. An electron gun at the top of a TEM release electrons that travel through the microscope vacuum tube. Rather than having a glass lens focusing the light, the TEM used an electromagnetic lens which focuses the electrons into a very fine beam. This beam then passes through the specimen and electrons scatter at the bottom of the microscope. An image of the sample with its assorted parts shown in different shades according to its density appears on the screen.

Equipment

- ☆ Eppendorf tube
- ☆ Vacuum drying oven
- ☆ Knife
- ☆ TEM
- ☆ Ultra-cut

Materials and Reagents

1. Trypsin working solution- 0.5 per cent trypsin and 0.2 per cent EDTA
2. Alcohol
3. 0.1 M phosphate buffer (PB)- 19 mL 0.2M NaH_2PO_4 and 81 mL 0.2 M Na_2HPO_4
4. Syringe
5. Dropper
6. Paraformaldehyde 4 per cent
7. Osimium tetraoxide 1 per cent
8. Epon- use syringe to mix all reagents to make epon and stored in 4°C

   Poly 812- 10 mL

   Araldite 502- 6 mL

   DDSA- 18 m

   DMP- 30.7 mL
9. Toluidine blue

   Toluidine blue 0.5 per cent

   Sodium borate 1 per cent

   Filter through 0.22μM filter
10. Lead citrate

    1 per cent lead citrate in water

    Add 0.5 mL of 10 M NaOH
11. Uranyl acetate
12. Propylene oxide (Use very carefully)

Procedure

1. After completion of treatment, cells treated with 0.5 per cent trypsin-EDTA, washing the cell through 0.5 mL 0.1 M phosphate buffer and then centrifuge at 300 rpm for 5 min.
2. Add 0.5 mL paraformaldehyde 4 per cent to fix cells at normal room temperature for 30 min and three times washing cell through 0.5 mL 0.1 M phosphate buffer.
3. Add 0.5 mL Osimium tetraoxide 1 per cent into the tube and allow to incubate for 1 h at room temperature.
4. Wash the cells three times using 0.5 mL 0.1 M phosphate buffer and then centrifuge at 300 rpm for 5 min.
5. Dehydrate cells by adding 0.5 mL 70 per cent alcohol for 5 min, centrifuge at 300 rpm for 5 min and remove alcohol.
6. Dehydrate cells by adding 0.5 mL 85 per cent alcohol for 5 min, centrifuge at 300 rpm for 5 min and remove alcohol.

7. Dehydrate cells by adding 0.5 mL 95 per cent alcohol for 5 min, centrifuge at 300 rpm for 5 min and remove alcohol.
8. Dehydrate cells by adding 0.5 mL 100 per cent alcohol for 5 min, centrifuge at 300 rpm for 5 min and remove alcohol.
9. Add 0.5 mL propylene oxide into new tubes and incubate for 3 min at room temperature, centrifuge at 300 rpm for 5 min and then remove propylene oxide.
10. Add 1 mL (propylene oxide: Epon, 1:1 ratio) into new fresh tubes and incubate for 1 h at room temperature, centrifuge at 300 rpm for 5 min and then discard (propylene oxide: Epon.
11. Add 1 mL Epon into fresh tubes, put the samples into the vacuum drying oven to vacuum for 1 h at room temperature and discard Epon.
12. Put tubes at 60°C oven for 48 h.
13. Trimming- remove the edge of the block.
14. Cut fine section of the samples (1μm) using sharp knife and transferred the sections on a slide and stain with 0.5 per cent toluidine blue for 40 second to 1 min and washed through sterile distilled water.
15. Put the thin sections on the grids and stain the sections through uranyl acetate and lead citrate for 3 min.
16. Wash the grids using sterile distilled water for 1 min and then allowing to complete drying.
17. Samples was ready to TEM images.

Chapter 10

Sodium Dodecyl Sulphate-Polyacrylamide Gel Electrophoresis (SDS-PAGE)

Introduction

Sodium Dodecyl Sulphate-Polyacrylamide Gel Electrophoresis (SDS-PAGE) is a discontinuous electrophoresis system produced by U.K. Laemmli, mostly used as a method to separate proteins according to their molecular weight.

Principle

SDS-PAGE is most commonly used gel electrophoresis system for estimation of the purity of protein in complex samples. In this technique, charged molecules in an electric field are migrated towards an electrode with opposite sign. The migration of molecules basically depends on various factors like net charge, voltage gradient of the electric field, shape and size of the molecule, frictional resistance of the supporting medium which impedes their movement. The basic principle of protein electrophoresis is, proteins have a net charge at any pH other than isoelectric point, and when transfer in an electric field, proteins molecules will migrate towards the electrode of the opposite charge. Basically, SDS is an anionic detergent and have negative charge which binds strongly to and denatures proteins to produce linear polypeptide chains. On average one molecule of SDS represent every two amino acids. In the presence of β- mercaptoethanol, help for protein denaturation by decreasing the disulphide bonds. The SDS mostly binds to the hydrophobic region of the denature protein. Negative charge nature of SDS damages the secondary and tertiary structure of proteins and strongly attracted toward a node in an electric platform. Because the charge-to-mass ratio is mostly the same among SDS- denatured proteins, the final separation of proteins is mostly depending on the differences in relative molecular weight of polypeptides. After visualization of protein specific bands done by staining process, the size of protein can be assessed by comparing its migration distance of proteins with ladder (molecular weight marker).

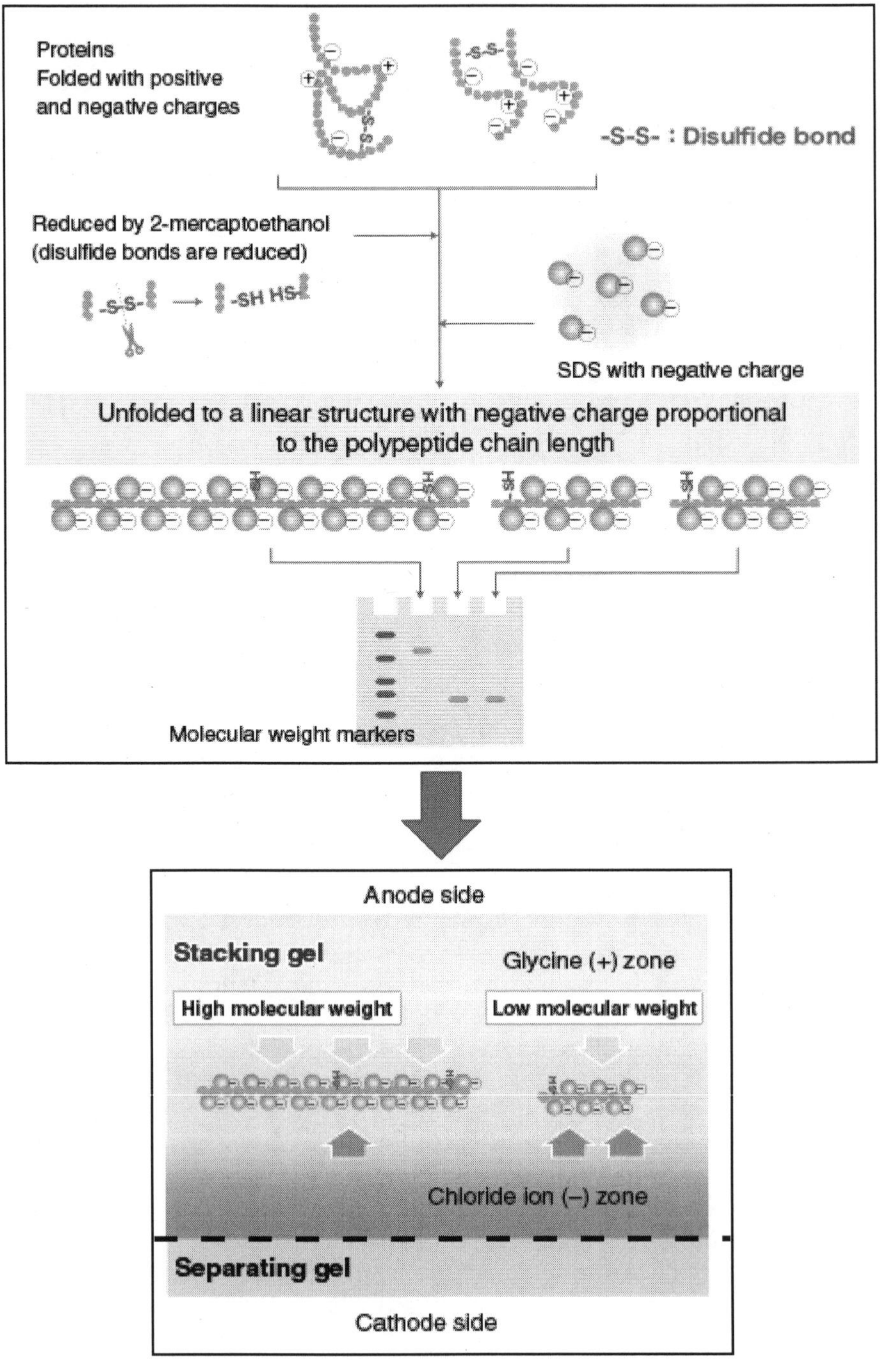

Material Required

1. **Power Supplies:** Convert the AC current into DC current.
2. **Gels:** These are either prepared in the laboratory or precast gels are purchased from the market.
3. **Electrophoresis Chambers:** The chambers that can fit the SDS-PAGE gels should be used.
4. **Protein Samples:** The protein is diluted using SDS-PAGE sample buffer and boiled for 10 minutes. A reducing agent such as dithiothreitol or β-mercaptoethanol is also added to reduce the disulfide linkages to prevent any tertiary protein folding.
5. **Running Buffer:** The protein samples loaded on the gel are run in SDS-PAGE running buffer.
6. **Staining and Destaining Buffer:** The gel is stained with Coomassie Brilliant Blue R- 250 Stain Solution. The gel is then destained with the destaining solution. Protein bands are then visible under naked eyes.
7. **Protein Ladder:** A reference protein ladder is used to determine the location of the protein of interest, based on the molecular size.
8. **Glasswares and others:** Conical flask, Beaker,
9. Measuring cylinder
10. Standard Protein molecular weight markers.
11. Micropipette for loading samples.

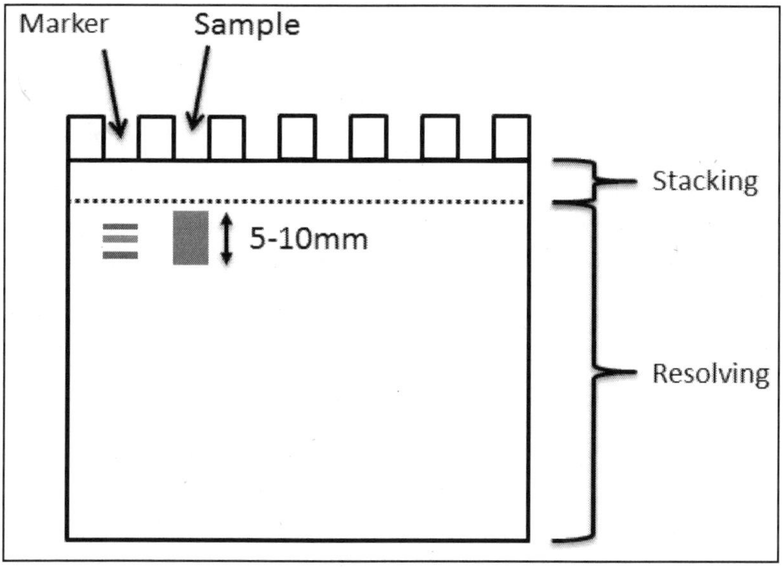

Gel Preparation Resolving Gel

Clean the glass plates and spacers of the gel casting unit with deionized water and ethanol.

Gel Per cent	*Distilled Water (mL)*	*30 per cent Acrylamide (mL)*	*1.5 M Tris-HCl, pH 8.8 (mL)*	*10 per cent SDS(μl)*	*10 per cent APS(μl)*	*TEMED (μl)*
8 per cent	4.6	2.6	2.6	100	100	10
10 per cent	3.8	3.4	2.6	100	100	10
12 per cent	3.2	4.0	2.6	100	100	10
15 per cent	2.2	5.0	2.6	100	100	10

Note: TEMED must be the last ingredient added

- ☆ Pour the gel solution in the plates assembled with spacers.
- ☆ Allow the gel to set for about 20-30 min at room temperature

Stacking Gel (5 per cent)

1.	Distilled water (mL)	5.65
2.	30 per cent Acrylamide (mL)	1.65
3.	0.5 M Tris-HCl, pH 6.8 (mL)	2.5
4.	10 per cent SDS (μl)	100
5.	10 per cent APS (μl)	100
6.	TEMED (μl)	10

Note: TEMED must be the last ingredient added

- ☆ Discard the overlayed water or isopropanol on the resolving gel
- ☆ Add the 5 per cent stacking gel solution until it overflows. Insert the comb immediately ensuring no air bubbles are trapped in the gel or near the wells.
- ☆ Allow the gel to set for about 20-30 min at room temperature.

Running Buffer

1.	Tris-HCl	12 g
2.	Glycine	57.6 g
3.	SDS	2 g

Note: Make up to 2000 mL with distilled water. Do not adjust pH with acid or base because pHis normally 8.3 as prepared.

Gel Loading Buffer/Sample Buffer

1.	1 M Tris-HCl (pH 6.8)	5mL
2.	10 per cent SDS	0.5 g
3.	Glycerol	4mL
4.	β-mercaptoethanol	0.25 mL
5.	0.5 M EDTA	1mL
6.	Bromophenol blue (0.5 per cent stock)	5mL

- Make up to 50 mL with distilled water.

Staining Solution

1. *Coomassie Brilliant Blue R-250*

- Staining of protein gels with Coomassie Brilliant Blue R-250 is a common procedure to visualize proteins resolved by SDS-PAGE. It is highly sensitive and is suitable for long-term storage of the gels.
- Take 0.1 per cent Coomassie brilliant blue R250 in 50 per cent methanol, 10 per cent glacial acetic acid. Dissolve the dye in the methanol and water component first, and then adds the acetic acid. Filter the solution through Whatman filter paper.
- **Note:** Coomassie brilliant blue is harmful by inhalation or ingestion. Wear appropriate gloves and safety glasses while handling.

Destaining Solution

- Take 10 per cent methanol and 7 per cent glacial acetic acid well mix using magnetic stirrer and stored.

Sample Preparation

- To a volume of protein sample, add equal volume of loading buffer.
- Boil the above mixture at 95°C for 5 min and centrifuge at 16000 rpm for 5 min.
- These samples can be stored at -20°C or may be used to proceed with gel electrophoresis.

Procedure

- Clean the internal surfaces of the gel plates with methylated spirits, dry, and then join the gel plates together to form the cassette, clamp it in a vertical position.

Casting the Gel/Preparation of Gel

Prepare Resolving Gel/Separating Gel

1. Firstly, fill the holder with water to check the leakage after 5 min. After leakage check, pour all water from gel casts and dry with the help of paper towel.
2. Prepare the separating gel solution by adding desired composition of reagents.
3. Two reagents namely TEMED and APS add in last, after solidification of the gel and then these reagents are added.
4. Mix the gel into gel casts up to a height approximately 1 cm below the gels comb.
5. Fill the remaining cast with isopropanol of distilled water on top of the gel.
6. Allow the gel to harden for 30-60 min.
7. Pour all isopropanol or distilled water from the gel casts and dry by the paper towel.

Prepare Stacking Gel

1. Mix the all reagents (except TEMED and APS) in proper amount to receive a gel with desired composition.
2. Add TEMED and APS in last just before pouring or after solidification of the gel.
3. Pipette gel mixture on top of separating gel to fill cast and remove all air bubbles.
4. Fill the remaining cast with isopropanol of distilled water on top of the gel.
5. Place the comb whilst the gel is still in liquid form. The comb will create wells to load the samples when the gel turns solid.
6. Allow the gel to harden for 20-30 min.
7. Drain the isopropanol or distilled water with strips of filter paper.

Running the Gel

1. Firstly, we confirmed the plates are secured and then mount the gels into the gel running tank.
2. Remove combs carefully (without any damage) and then fill the inner chamber of the tank with the help of running buffer to the top.
3. After filling inner chamber of the tank, fill a third of the volume in the outer chamber of the tank with running buffer.
4. Rinse the loading tip a few times with distilled water.

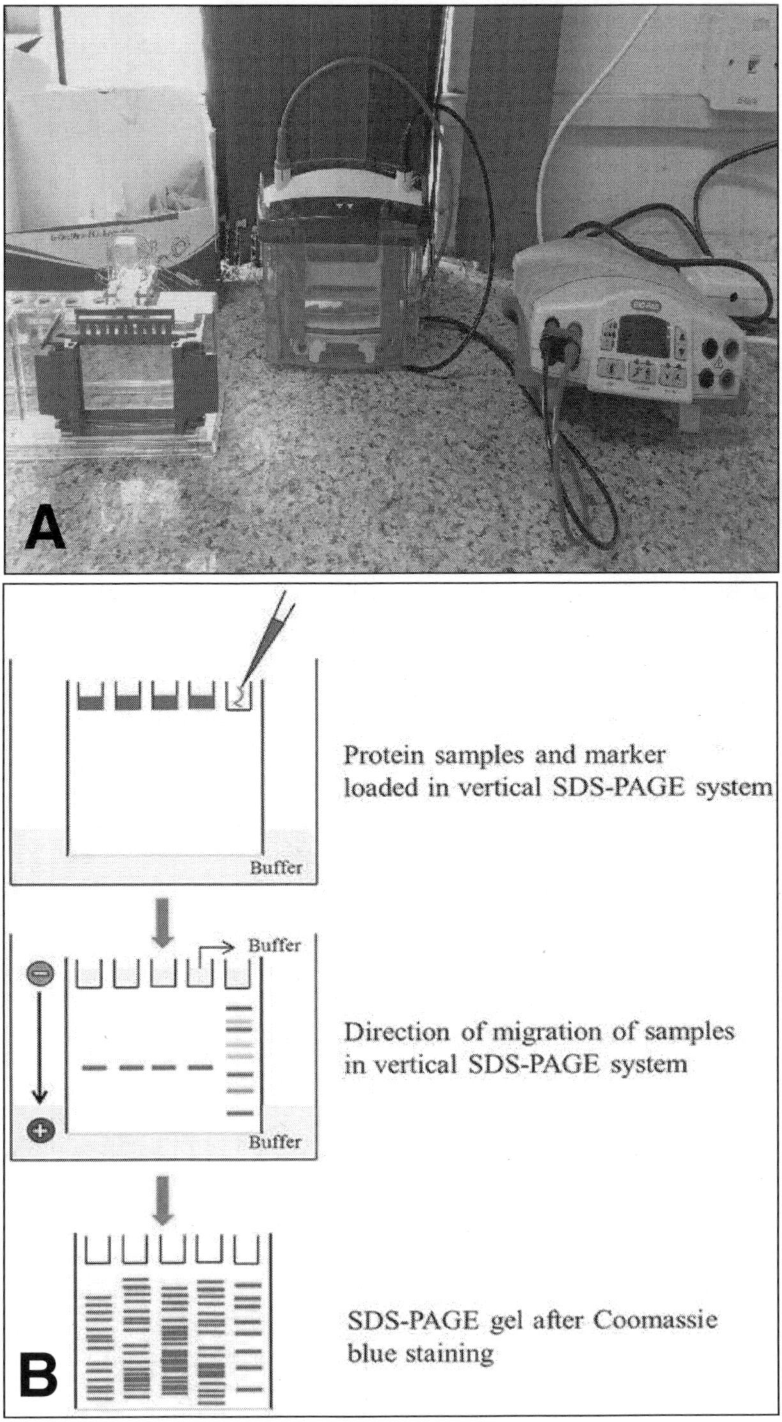

Figure 10.1: A. Set up of SDS-PAGE,
B. Protein samples and color burst protein marker.

5. Insert the loading tip to a few mm from the well bottom and load 2μl ladder with the help of pipette and then deliver the 10 μl sample into each well.
6. Connect power supply by putting the lid (red-red, black-black) and run at 100 V for 60 min or until the blue line almost reaches the bottom of the gel.
7. After completion of electrophoresis, remove gels from the glass plates.

Staining Procedure

After the electrophoresis, place the gel in a plastic tray containing gel fix solution. Place the tray on a rocking table and fix the proteins for 2 hours. Remove the gel fix solution and add Coomassie blue staining solution and stain the gel for 2-4 hours. After the completion of staining process, wash the gel 4-5 times with distilled water to remove excess stain. Add destain solution to the gel and kept for 4 hours till clear blue bands on clear background are visible. After destaining, the gels may be stored in gel storage solution and photographed as required. Calculate the molecular weight of the protein bands with the help of known molecular weight ladder (markers).

Note

1. β-mercaptoethanol is essential for disrupting disulphide bridges in proteins. However, exposure to air decreases the reducing power of β-mercaptoethanol. Thus, it should be prepared fresh.
2. Destain solution needs to be replaced at regular intervals since a simple equilibrium is quickly set up between the concentration of stain in the gel and destain solution after which no further destaining takes place.
3. It is generally accepted that a very faint protein band detected by Coomassie brilliant blue, is equivalent to about 0.1μg(100ng) of protein.
4. Data Analysis Label each lane on the photograph of your gel: The molecular weight of the unknown protein can be determined by running calibration proteins of known molecular weight on the same gel run as the unknown protein.

Chapter 11

Native-Polyacrylamide Gel Electrophoresis (PAGE)

Introduction

Native PAGE, also known as non-denaturing PAGE, is run in the absence of SDS and also used for separation of the native protein. This method can be following the standard protocol described by Laemmli SDS-PAGE protocol, except the use of SDS or reducing agent. In native PAGE, separation of protein is mainly based on size, native charge and shape, and also determines the homogeneity of the purified protein. In SDS-PAGE, denatured proteins are basically negative charged and electrophoretic mobility of proteins mainly depends on their molecular weights, but in case of native PAGE, the electrophoretic mobility depends on the charge and hydrodynamic size of the protein.

Principle of Native-PAGE

Native PAGE mainly uses same discontinuous chloride and glycine ion fronts as SDS-PAGE to create moving boundaries that stack and then separate proteins by their charge to weight ratio. Proteins are prepared in a native-PAGE sample buffer, which also helps to maintain the protein secondary structure and their native charge density. When your target protein has polymerized, you can easily see multiple bands from your native PAGE gel. In native PAGE mostly proteins are acidic or slightly basic pH nature and migrate towards the negative polar.

Materials

- ☆ Gel loading tips for micropipette
- ☆ Measuring cylinder
- ☆ Rocking shaker

- ☆ Eppendorf tube
- ☆ Power supply
- ☆ Water bath
- ☆ Gel dryer
- ☆ Gel apparatus

Reagents

1. Acrylamide
2. Tris
3. Hydrochloric acid (HCl)
4. TEMED (N,N,N′,N′-tetramethylethylenediamine)
5. Ammonium persulfate
6. Glycine
7. Glycerol
8. Bromophenol blue

Reagent Preparation

1. Native PAGE Stacking Gel (5 per cent, 2 mL)

0.5 M Tris-HCl pH 6.8	500 µl
Acrylamide (30 per cent)	333 µl
Ammonium persulfate (10 per cent)	20 µl
TEMED	2 µl
Distilled water	1.125 mL

Note: TEMED must be the last ingredient added

2. Native PAGE Separating gel (12 per cent, 5 mL)

Acrylamide (30 per cent)	2 mL
1.5 M Tris-HCl pH 8.8	1.25 mL
Ammonium persulfate (10 per cent)	50 µl
TEMED	5 µl
Distilled water	1.65 mL

Note: TEMED must be the last ingredient added

3. Sample Buffer (5X, 10 mL)

1 M Tris-HCl (pH 6.8)	3.1 mL
50 per cent Glycerol	5mL
1 per cent Bromophenol blue	0.5 mL
Distilled water	1.4 mL

4. *Running Buffer/Electrophoresis Buffer (1 liter)*

25 mM Tris base	3 g
192 mM Glycine	14.4 g
Distilled water	Make up 1 litre

5. **Ammonium persulfate solution (10 per cent, 5 mL)-** Dissolve 0.5 gm ammonium persulfate in 5 mL distilled water
6. **Acrylamide solution (30 per cent)-** Dissolve 29.2 gm acrylamide with 100 mL distilled water.
7. **Bromophenol blue solution (1 per cent, 10 mL)-** Weight 100 mg bromophenol blue and add10 mL distilled water.
8. **2 M Tris-HCl pH 8.8 (final con. 1.5M)-** For 1 litre, weight 242 gm Tris base and add 500 mL distilled water. Adjust pH by adding concentrated HCl and finally add distilled water to make up total volume to 1 litre.
9. **1 M Tris-HCl pH 6.8 (final con. 0.5M)-** For 100 mL, weight 12.1 gm Tris base and add 50 mL distilled water. Adjust pH by adding concentrated HCl and finally add distilled water to make up total volume 100 mL.

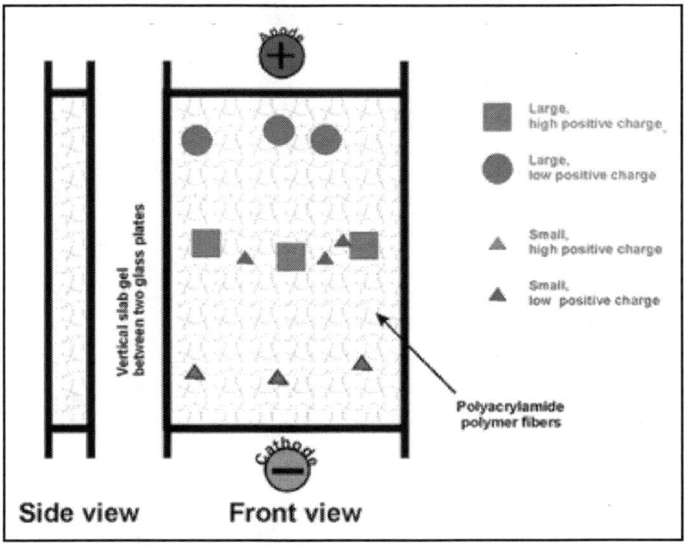

Figure 11.1: Movement of Molecules According to their Size and Charges in Native-Polyacrylamide Gel Electrophoresis (PAGE).

Procedure

1. Prepare appropriate amount of separating gel in a small beaker, then gently swirl the beaker to ensure a sufficient mixing. Pipette the gel solution into the gap between the glass plates of gel casting (Don't fully fill). Fill the

rest space with water (isopropanol alternatively). Allow 20-30min for a complete solidified.

2. Prepare the stacking gel solution while the separating gel is solidified. Prepare appropriate amount of stacking gel in a beaker and mix well and pour out the water in the first step and pipet the stacking gel solution into the gap and insert the comb. Allow 20-30 min to let it solidified.
3. Mix your sample with sample buffer and load the sample mixture into each well by loading tip to a few mm from the well bottom.
4. Connect power supply by putting the lid and run at 100 V for 60 min or until the blue line almost reaches the bottom of the gel.
5. After the electrophoresis, place the gel in a plastic tray and add Coomassie blue staining solution and stain the gel for 2-4 hours.
6. After the completion of staining process, wash the gel 4-5 times with distilled water to remove excess stain.

Chapter 12

Native-PAGE Analysis for Isoform Profile of PO, PPO, SOD, Catalase and Ascorbate Peroxidase

Introduction

The isoform profile of peroxidase (PO), polyphenol oxidase (PPO) and superoxide dismutase (SOD) were examined by non-denaturing native polyacrylamide gel electrophoresis (Native- PAGE).

1. Native Gel Activity of Peroxidase (PO) (Sindhu *et al.*, 1984)

Introduction

Native gel electrophoresis was carried out to study the expression pattern of various isoforms of POs.

Reagents

A. **Sample buffer-** 150 mM Tris-HCl (pH 6.8), 3 per cent Bromophenol blue, 30 per cent Glycerol

B. **Diamino benzidine (DAB) solution-** 0.15 per cent benzidine, 6 per cent NH4Cl

Procedure

1. The protein extract prepared by homogenizing 1 gm of potato leaves samples in 2 mL of 0.1M sodium phosphate buffer and then centrifuged at 16,000 rpm for 10 min at 4°C.

2. The concentration of protein in the sample was determined by Bradford method.
3. Samples (containing 50 μg protein) were mixed with sample buffer and separated by native-PAGE by using 12 per cent polyacrylamide gel.
4. After electrophoresis, the gel was equilibrated in 50 mL of 0.1 M potassium phosphate buffer (pH 5.0) for 30 min.
5. Peroxidase isoform were then visualized by incubating the gel in the dark condition with freshly prepared diamino benzidine (DAB) solution for 15-20 min.
6. Then add 10 μl of 30 per cent H_2O_2 and constant shaking till bands appear.
7. After staining, the gel must be washed with distilled water and take photographs.

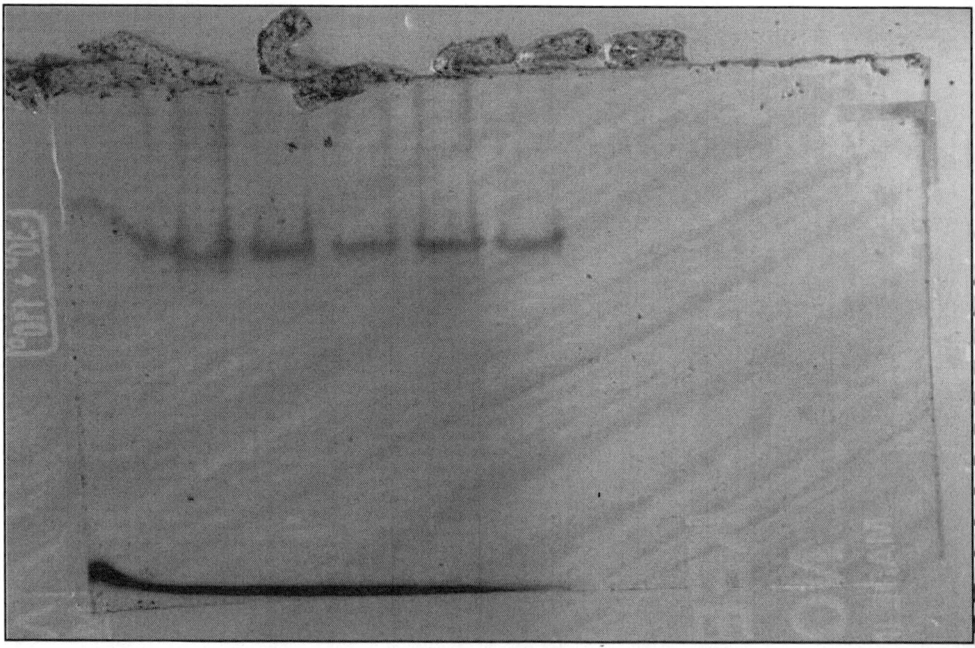

Figure 12.1: Expression of Peroxidase Isoforms

2. Native Gel Activity of Polyphenol Oxidase (PPO) (Jayaraman *et al.*, 1987)

Procedure

1. Take 1 gm of fresh potato leaves and homogenized in 2 mL of 0.01 M potassium phosphate buffer (pH 7.0).
2. The homogenate was centrifuged at 20,000 rpm for 15 min at 4°C and the supernatant used as enzyme source.

3. For native polyacrylamide gel electrophoresis, prepared 8 per cent resolving gel/separating gel and 4 per cent stacking gel.
4. After native gel electrophoresis, the gel was equilibrated in 0.1 per cent *p*-phenylene diamine in 0.1 M potassium phosphate buffer (pH 7.0) for 30 min.
5. Then add 10 mM catechol in the same buffer. After addition of catechol, constant shaking till the appearance of dark brown discrete bands.

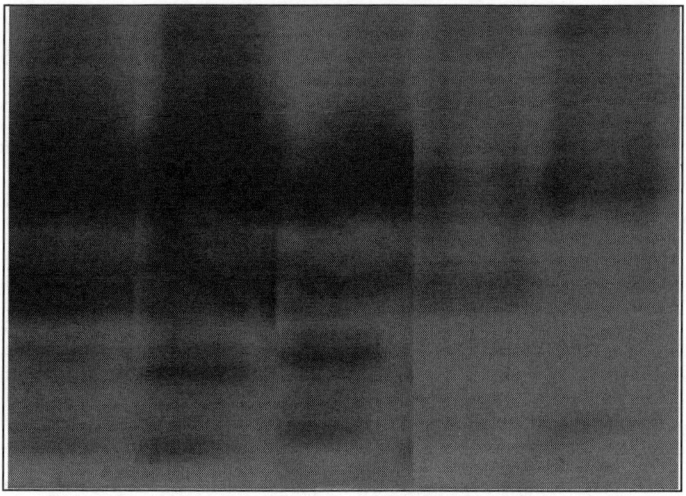

Figure 12.2: Expression of Polyphenol Oxidase Isoforms.

3. Native Gel Activity of Superoxide Dismutase (SOD) (Azevedo *et al.*, 1998)

Introduction

Native gel electrophoresis was carried out to study the expression pattern of various isoforms of superoxide dismutase.

Reagents

1. **Extraction buffer-** 50 mM sodium phosphate buffer (pH 7.4) with the addition 1mM EDTA and 1 per cent polyvinylpyrrolidone (w/v).

Procedure

1. Extract prepared from 1 g fresh potato leaves homogenized with a pre-chilled mortar and pestle in 3 mL extraction buffer.
2. Then centrifuged at 10,000 rpm for 20 min at 4°C and supernatant used as enzyme source.
3. Native electrophoresis was carried out under 8 per cent polyacrylamide gel for SOD activity.

4. Sample were mixed with sample buffer and loaded on to each lane.
5. After native electrophoresis, the gels were rinsed in distilled water and incubated in the dark condition at room temperature for 30 min in the stain solution (Tris 0.61 gm; water 100 mL; Sodium EDTA 7.5 mg; Riboflavin 4 mg and nitroblue tetrazolium 10 mg).
6. At the end of this process, the gel placed in distilled water and exposed on a light box for 10 min until the appearance of colourless bands in a purple gel was visible.
7. The reaction stopped by placing the gels to 6 per cent acetic acid.

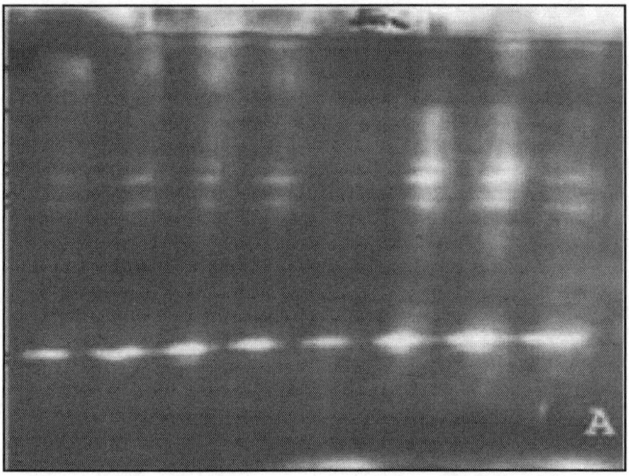

Figure 12.3: Expression of Superoxide Dismutase Isoforms.

4. Native Gel Activity of Catalase (CAT) (Woodbury *et al.*, 1971)

Reagents

1. **Extraction buffer-** 50 mM sodium phosphate buffer (pH 7.4) with the addition 1mM EDTA and 1 per cent polyvinylpyrrolidone (w/v).

Procedure

1. Prepare extract from 1 g fresh potato leaves homogenized with a pre-chilled mortar and pestle in 3 mL extraction buffer.
2. Then centrifuge at 10,000 rpm for 20 min at 4°C and supernatant used as enzyme source.
3. For visualization of CAT isoforms, used 10 per cent non-denaturing polyacrylamide gel for catalase activity.
4. Sample are mixed with sample buffer and loaded on to each lane.
5. After native electrophoresis, gel is rinsed in distilled water for 15 min and then incubated in 0.003 per cent (v/v) H_2O_2 for 10 min and rinsed with distilled water.

6. Then gel is stained in 1 per cent ferric chloride and 1 per cent potassium ferricyanide solution with constant shaking to appearance the CAT isoforms.

   Note: Prepare the solution separately and mix fresh before using.

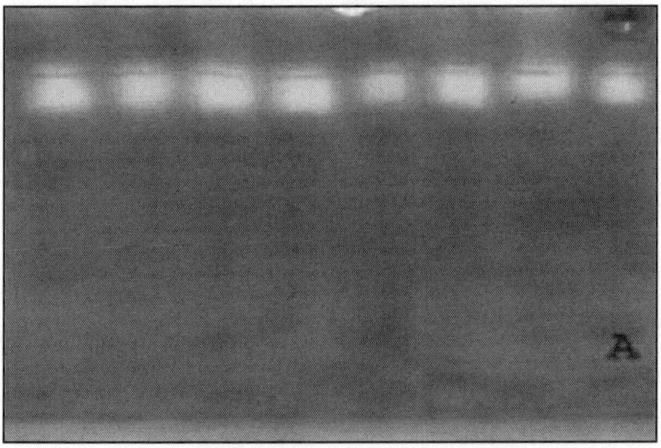

Figure 12.4: Expression of Catalase Isoforms.

5. Native Gel Activity of Ascorbate Peroxidase (APX) (Mittler and Zilinskas, 1992)

Reagents

1. **Extraction buffer-** 50 mM sodium phosphate buffer (pH 7.4) with the addition 1mM EDTA and 1 per cent polyvinylpyrrolidone (w/v).

Procedure

1. Prepare the extract from 1 gm fresh potato leaves homogenized with a pre-chilled mortar and pestle in 3 mL extraction buffer.
2. Then centrifuge at 10,000 rpm for 20 min at 4°C and use the supernatant as enzyme source.
3. Carry out native electrophoresis under 10 per cent non-denaturing polyacrylamide gel for ascorbate peroxidase activity.
4. The carrier buffer comprises 2 mM ascorbate and pre-run the gel for 30 min to allow ascorbate to enter the gel.
5. Mix the sample with sample buffer and load on to each lane.
6. After native gel electrophoresis, immerse the gel in 50 mM sodium phosphate buffer (pH 7.0) containing 2 mM ascorbate for 30 min.
7. Then incubate in 50 mM sodium phosphate buffer (pH 7.0) comprised 4 mM ascorbate and 2 mM hydrogen peroxide (H_2O_2) for 20 min.

8. Wash the gel with the buffer for 1 to 2 min and then incubate in a 50 mM sodium phosphate buffer (pH 7.8) containing 28 mM TEMED and 2.45 mM nitroblue tetrazolium.
9. After staining, wash the gel with distilled water and take photographs.

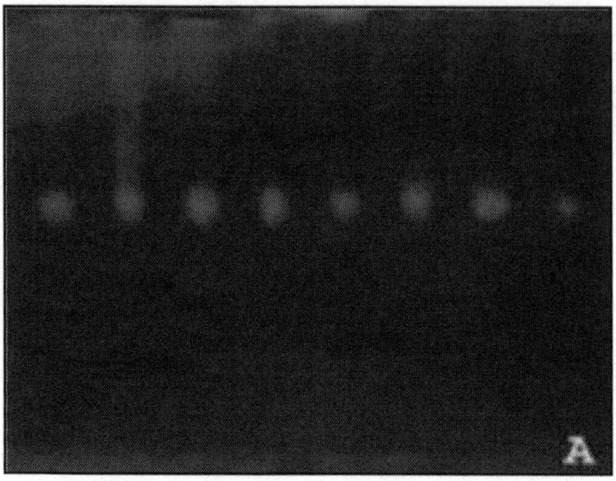

Figure 12.5: Expression of Ascorbate Peroxidase Isoforms.

Chapter 13

Plant Genomic DNA Extraction Using CTAB Method

Introduction

Isolation and purification of DNA is a pivotal process in molecular biology techniques used in plant studies for the nature of genotypes, economical traits associated with gene of interests (GoI). Isolating DNA from the plant materials can be very challenging to achieve high quality DNA, because having variable levels of secondary metabolites and structural biomolecules. Polysaccharides and polyphenols are two most prominent classes of plant biomolecules and cause problems in extraction of plant DNA. These biomolecules also affect the process of DNA purification. These plant components have a similar structure of nucleic acids that allow variable metabolites, polyphenols and polysaccharides to disturb the total DNA isolation. They also combined with nucleic acid during DNA extraction and also affect the quality and quantity of the isolated DNA from the plants, and purity of the isolated nucleic acids.

Currently, various extraction protocols have been established to solve the polysaccharides and polyphenols problems, and isolate a good quality and intact whole genomic DNA from plant tissues. Among them, use of CTAB (Cetyltrimethylammonium bromide), a cationic detergent, facilitates the separation of polysaccharides during purification while additives, such aspolyvinylpyrrolidone, can aid in removing polyphenols. CTAB based extraction buffers are widely used when purifying DNA from plant tissues.

Materials Required

- ☆ CTAB buffer
- ☆ Chilled mortar and pestle

- ☆ Liquid nitrogen
- ☆ Tubes
- ☆ Ethanol 70 per cent
- ☆ Ammonium acetate
- ☆ Chloroform: Isoamyl alcohol (24:1)
- ☆ Distilled water

Reagents

1. *CTAB Buffer*

CTAB	2 gm
Tris base 1 M (pH 8.0)	10.0 mL
EDTA 0.5 M (pH 8.0)	4.0 mL
NaCl 5 M	28.0 mL
PVP 40	1 gm
Distilled water	40.0 mL

Note: Make up 100 mL volume by distilled water

Tris Base 1 M (pH 8.0)

Dissolve 121.1 g of Tris base in 800 mL of distilled water and adjust pH to 8.0 by adding 42 mL of concentrated HCl. Allow the solution to cool at room temperature before making the final adjustments to the pH. Adjust the volume to 1 L with distilled water and then sterilize by using an autoclave.

Procedure

1. Take 200 mg of plant tissue and grind with 500 µl of CTAB buffer and centrifuge at 12000 g for 5 min to spin down cell debris. Transfer the supernatant to clean centrifuged tubes.
2. To each tube add 250 µl of Chloroform: Isoamyl alcohol (24:1) and mix the solution by inversion. After mixing, centrifuge tubes at 13000 rpm for 1 min and then transfer the upper aqueous phase only (contains the DNA) to a clean tube.
3. Add 50 µl of 7.5 M ammonium acetate followed by 500 µl of absolute ethanol and invert the tubes several times to precipitate the DNA.
4. Generally, the DNA can be seen to precipitate out of solution. Alternatively, the tubes can be placed for 1 hr at -20°C after the addition of ethanol to precipitate the DNA.
5. To wash the DNA, transfer the precipitate into a microfuge tube containing 500 µl of 70 per cent chilled ethanol and slowly invert the tube.

6. After the wash, spin the DNA into a pellet by centrifuging at 13000 rpm for 1 min. Remove all the supernatant and allow the DNA pellet to dry for 15 min.
7. Resuspend the DNA in sterile DNase free water (approximately 50-400 µl H_2O; the amount of water needed to dissolve the DNA can vary, depending on how much is isolated). RNaseA (10 µg/mL) can be added to the water prior to dissolving the DNA toremove any RNA in the preparation (10 µl RNaseA in 10mL H_2O).
8. After resuspension, the DNA is incubated at 65°C for 20 min to destroy any DNases that may be present and store at 4°C.

Chapter 14

DNA Quantification

Introduction

DNA can be quantifying through different methods. Some are popular and we have summarized here.

1. UV Absorbance

Introduction

This method is the most common method for quantification of DNA. The absorbance of were measured by apparatus when light passes through the liquid DNA material. DNA will absorb at specific wavelengths of light. The Nanodrop spectrophotometer is one of the instruments designed to absorbance of DNA when extracted sample keep on the tip of the absorbance point.

Procedure

1. First, we have to measure the blank. The blank comprises with elusion buffer where we have diluted the DNA.
2. Then we measure the absorbance of DNA. The absorbance will give the idea of concentration of DNA in the preparation. And also gave information about the contaminants in the DNA.
3. DNA and RNA maximum absorbed at 260 nm proteins at 280 nm and other impurities like organic compounds and chaotropic will absorb at 230 nm. The A260/A280 nm is the indication of DNA purity. Ideally the number should be between 1.8 to 2.0 and A260/A230 should be 1.5.
4. Then, using the A260 reading, you can calculate the DNA concentration. Generally, A260 of 1.0 is equivalent to 50 ug/mL pure dsDNA. Use the following formula to estimate DNA:

   Concentration (ug/mL) = A260 reading x dilution factor x 50 ug/mL

2. Agarose Gel Electrophoresis

Procedure

1. Generally used the 0.8 per cent gel for running the DNA in to the gel electrophoresis unit for knowing the DNA condition either DNA is intact or in correct size.
2. The gel has been prepared and after setting the comb in the gel tray we used to load the ladder than extracted DNA in to next well.
3. The gel will run the band intensity will be monitored and compare with the ladder which is closest to the DNA band.
4. This method is best suited for PCR product amplification. This band also gave the contamination presence through indication of other bands during the gel run.
5. The thicker the band at gel pictures the more concentration of DNA. According to that we will dilute our DNA.

Chapter 15

RNA Isolation

Introduction

The isolation of a good quality of RNA from plant tissues is the primary requisite for studies in plant science. Plant tissues contain huge number of polysaccharides, phenolic compounds and different types of secondary metabolites, which makes it difficult to extract a high and good quality RNA from them. For isolation of high-quality RNA, too much skilled person is required because ribonucleases (RNases) are very sensitive and active. There are a number of methods described for the extraction of total RNA and the method selected largely depends upon the source material and the experimental purpose for which RNA is extracted. Various methods are available for isolation of RNA.

1. Phenol-chloroform method (Brawerman, 1974)
2. Cell lysis method (Zhang *et al.*, 2018)
3. TRIZOL method (Rio *et al.*, 2010)

1. Phenol-Chloroform Method

Principle

The ribonucleoprotein complex is dissociated by SDS into RNA and protein, deproteinized by phenol and the free RNA left in aqueous solution is precipitated in the cold after adding alcohol.

Materials Required

- ☆ Phenol
- ☆ Extraction buffer
- ☆ Cold condition (4°C)
- ☆ Magnetic stirrer

- ☆ Ethanol
- ☆ SDS 10 per cent (w/v)
- ☆ Ether

Reagents

1. *Prepare Extraction Buffer*

Tris-HCl (0.1 M) – 1.21 g

NaCl (0.075 M) – 0.44 g EDTA Na_2 (0.005 M) – 0.19 g

Water to – 100 mL

Procedure

1. Take 0.5-5 g leaf sample and grind in a pre-chilled mortar and pestle with the help of liquid N_2, fine powder extract using 10 mL of extraction buffer and then centrifuge at 2,000 rpm for 3 min.
2. After centrifugation, transfer the supernatant to a volumetric flask and stir with 0.1 volume of 10 per cent SDS for 2–3 min.
3. Add an equal amount of buffered phenol (freshly redistilled phenol saturated overnight with 100 mM Tris-HCl, pH 8.5).
4. Partition the content by centrifuging at 2,000 rpm for 5 min and collect the upper aqueous part into a separate sterilized flask.
5. Shake the lower and interphase again with an equal amount of extraction buffer for 5 min and again centrifuge.
6. Combine the aqueous phase with the step 4 and stir with an equal amount of buffered- phenol for 5 min.
7. Repeat the extraction and centrifugation step four to five times or until the interphase shows no proteins.
8. Finally, collect the upper aqueous parts containing RNA and dissolve in it about 250 mg NaCl.
9. Add cold ethanol (96 per cent) and leave the flask overnight at 20°C for RNA precipitation.
10. Collect RNA by centrifugation at 2,000 rpm for 10 min. Wash the pellet (RNA) by using 70 per cent ethanol, ethanol: ether (1:1 v/v). Gently dry the pellet *in vacuo* for a few minutes.
11. RNA completely dissolved in elution buffer for further analysis.

2. Cell Lysis Method

Introduction

This method uses the lysis buffer and this protocol eliminates the need of binding and elution from solid surfaces.

Materials Required

- ☆ Mortar and pestle
- ☆ Stainless spatula
- ☆ Centrifuge tubes (50 mL)
- ☆ Pipettes
- ☆ Vortex
- ☆ Disposable tips and gloves
- ☆ Centrifuge machine
- ☆ Eppendorf tubes
- ☆ For RNases free experiment: All tips and tubes soaked in 0.1 per cent DEPC for overnight at room temperature and then autoclaved. The stainless spatula and mortar and pestle baked at 180°C for 12 h.

Reagents

- ☆ Extraction buffer (0.2 M Tris, 0.4 M KCl, 0.2 M sucrose, 35 mM $MgCl_2$, 25mM EGTA)
- ☆ Phenol (pH 8.0)
- ☆ Chloroform: isoamyl alcohol (24:1, v/v)
- ☆ 3 M NaAc (pH 5.2)
- ☆ 3 M NaAc (pH 5.6)
- ☆ 0.3 M NaAc (pH 5.6)
- ☆ DEPC treated water

Procedure

This procedure can be applied to a wide range of herbaceous plants.

1. Take 1 gm of fresh leaf sample and grind the sample with liquid nitrogen in a pre- chilled mortar and pestle to a fine powder.
2. Immediately transfer the powder centrifuged tube (50 mL capacity), then add phenol (2 mL), then add 4 mL extraction buffer and then 2 mL chloroform: isoamyl alcohol (24:1) ratio were added subsequently.

   Note: Phenol is added first that creates denaturation conditions.
3. Vortex the centrifuged tube until the complete emulsion was formed and then centrifuged at 8,000 rpm for 5 min at 4°C.
4. Then transfer upper aqueous part to another 50 mL centrifuged tube, and add 2 mL phenol and 2 mL chloroform: isoamyl alcohol (24:1) mix properly and centrifuged at 8,000 rpm for 5 min at 4°C.
5. Take the upper part and transfer another 50 mL tube and add 4 mL chloroform: isoamyl alcohol (24:1) then mix properly and centrifuged sample as before.

6. Transfer upper part to another 50 mL tube and record the volume that was transferred.
7. After transfer, add 1/10 volume of 3 M NaAc (pH 5.2) and add 2.2 volume of ethanol and mix it by inversion several times.
8. After embedding the tubes, precipitate in crushed ice for 10 min and then centrifuged at 15,000 rpm for 15 min at 4°C.
9. Resuspend the nucleic acid sediments by adding 1 mL of 3 M NaAc (pH 5.6) with the help of pipette and centrifuged at 15,000 rpm for 15 min at 4°C.
10. Very carefully discard the supernatant and place the tube invertedly on a filter paper to allow the air-dry the nucleic acid.
11. Resuspend the RNA pellets in 0.4 mL 0.3 M NaAc (pH 5.6) and then add 1 mL of ethanol, mix the tubes several times.
12. Precipitate in crushed ice for 10 min and then centrifuged at 15,000 rpm for 15 min at 4°C.
13. Then wash the RNA pellet by using 0.2 mL of 70 per cent ethanol, then air-dry and then dissolve the pellet in 0.5 mL DEPC water.

3. TRIZOL Method for RNA Extraction

Introduction

Trizol solubilization and extraction is recently developed method for deproteinizing RNA and also known as TRI Reagent used for isolation of total RNA. Basically, Trizol is a potential monophasic solution of guanidine thiocyanate and phenol, which effectively dissolves the protein, DNA and RNA on homogenization of sample. After adding the reagent such as chloroform and centrifugation process, can be separates mixture into 3 parts and upper most layer or parts containing the RNA. Therefore, protein, DNA and RNA can be purified from a single sample. Trizol method is basically used for isolating small RNAs, that is, micro RNAs small interfering RNAs and piwi-associated RNA.

Materials

- Trizol reagent
- PCI
- Ethanol
- Chloroform
- Isopropanol
- Eppendorf tube (1.5- and 2.0-mL size)
- Centrifuged tube (5 mL)
- DEPC water
- Vortex mixer
- Water bath

- ☆ DNA-free kit
- ☆ Pipettes
- ☆ Magnetic stir
- ☆ Dry ice

Procedure

1. Sample preparation (1 mL Trizol/50-100 mg tissue) and homogenized.
2. Transfer the fine powder to 50 mL centrifuged tube and vortex mixture thoroughly.
3. Aliquot the solution to eppendorf tube and stand at room temperature for five min.
4. Add 0.2 mL chloroform per 1 mL of Trizol and shake vigorously for 15 seconds and thencentrifuged the samples at 12,000 rpm for 15 min at 4°C.
5. Centrifugation can separate the obtained mixture into three phases:
   a. Upper phase containing RNA
   b. Interphase containing DNA
   c. Lower phase containing protein

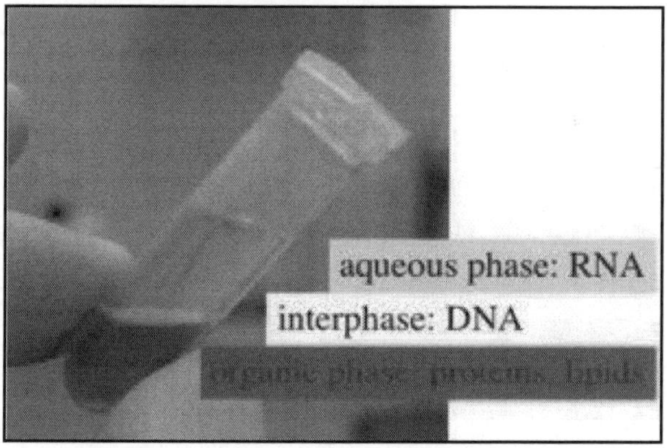

6. Transfer the upper aqueous phase to another tube.
7. Add 0.5 mL isopropanol per 1 mL of Trizol, mix properly by inverting the tube several times and incubate at room temperature for 15 min and then centrifuge at 10,000 rpm for 10 min at 4°C.
8. After centrifugation, remove the supernatant and wash RNA pellet by using 1 mL of 70 per cent ethanol per 1 mL of Trizol and then centrifuge the sample at 14,000 rpm for 15 min at 4°C.
9. Remove the supernatant and invert the tube, air-dry. Dry pellet incubates in dry bath at 55°C for 5 min.
10. Dissolve the pellet in 0.8 mL DEPC treated water.
11. To obtained DNA contamination free total RNA is treated by DNase (Ambions, DNA- free kit).

12. The kit comprised DNase enzyme and buffer that remove DNA from RNA suspensions.
13. Add 10 per cent of 10X DNase '! buffer (5µl for 50µl of RNA), then add 2µl of DNase '! enzyme and mix properly and incubate at 37°C for 20-30 min.

Chapter 16

Methods of RNA Quality Assessment

Introduction

Accurate determination of high-quality RNA is important for many downstream applications namely cloning, next generation sequencing (NGS) and transfection. The success of these downstream RNA based techniques is relied on obtaining a good quality RNA and it is compulsory to calculate the quantity and quality of purified RNA before setup the downstream assays. Improper quantification can enhance the variability in downstream process and can affect the quality of RNA sample results. RNA is very sensitive and easily susceptible to degradation due to pervasive presence of RNases in the natural environment. The successful isolation and analysis of RNA involves not only great care during extraction and handling but also involves selecting the best and appropriate purification and analysis methods. Therefore, the quality, purity and integrity of RNA should be checked by using various appropriate methods.

1. Agarose gel electrophoresis (Armstrong *et al.*, 2015)
2. RNA quantification using UV absorbance
3. Real time quantitative PCR
4. 2100 Bioanalyzer
5. Fluorescent dye-based quantification

1. Agarose Gel Electrophoresis

Introduction

One of the most valuable methods used to assess the integrity of total RNA analysis is to run an aliquot of the RNA sample on a denaturing agarose or acrylamide gel stained by using ethidium bromide (EtBr). In case using native (non-denaturing) gels, it is very problematic to interpret the data and different secondary structure of RNA can changed the migration pattern in native gels. Samples are loaded into the agarose gels, and power is passed through the gel, RNA fragments can separate based on the size. The gels containing various separated fragments are stained by using different types of fluorescent dye that bind with RNA, namely ethidium bromide (EtBr) and SYBR Green. This method is almost similar in many respects to DNA agarose gel electrophoresis but, major difference in the inclusion of formaldehyde in the RNA sample and in agarose gel. The role of formaldehyde is to denature the RNA to prevent development of regions of dsRNA, which affects the migration of the RNA.

Materials Required

- ✰ Erlenmeyer flask
- ✰ Water bath
- ✰ Power supply
- ✰ RNase free electrophoresis system (gel tray, comb and buffer chamber)
- ✰ Containers for staining and destaining gels
- ✰ Gel documentation system
- ✰ Microwave oven

Reagents

1. Agarose
2. Formaldehyde 37 per cent (12.3 M, pH 4.0)
3. RNase free distilled water
4. Ladder (RNA of known size)
5. 0.5 ammonium acetate
6. RNA loading dye, 6X
   10 mM EDTA
   Bromophenol blue 0.25 per cent (w/v)
   Xylene cyanol 0.25 per cent (w/v)
   Glycerol 50 per cent (w/v)
7. 10X MOPS running buffer
   0.2 M 3-(N-morpholino)- propanesulfonic acid (MOPS)
   0.1 M sodium acetate 10 M EDTA (pH 8.0)
8. 50 X TAE buffer (pH 8.0)
   Tris base: 242 g (0.89 M)
   Glacial acetic acid: 57.1 mL (0.89
   M)EDTA 0.5 M: 100 mL (0.02 M)
   ddH_2O- 1 L

Note: All solution should be prepared with DEPC treated sterile water (for RNA agarose gel electrophoresis) and used distilled water in all protocol steps.

Procedure

1. *Prepare and Pour the Agarose Gel*

(a) In an Erlenmeyer flask, add 1 gm agarose in 72 mL of distilled water and heat the solution in a microwave oven until completely dissolved, then allow the solution to cool to 60°C.

(b) Add 10 mL of 10X MOPS running buffer and then add 18 mL of 37 per cent formaldehyde (12.3 M).

Note: Formaldehyde is toxic and directly absorbed through the skin and inhalation of vapours.

(c) Setup the gel tray, pour the gel using comb that will create wells and join the gel in the gel tank and add 1X MOPS running buffer until the fully cover the gel and then remove the comb.

2. *Preparation of RNA Sample*

(a) Add RNase free water to each RNA sample and 1-3µg RNA should be loaded per lane.

(b) Load 0.5 X formaldehyde load dye, but the completely denaturate the RNA should be used 3X formaldehyde load dye.

(c) Add ethidium bromide to the formaldehyde load dye at a final concentration of 10µg/mL.

3. *Running the Denaturing RNA Gel*

(a) Run the gel at 5V/cm until the bromophenol blue dye has migrated at least 2-3 cm into the gel.

4. *Staining the RNA Agarose Gel with Ethidium Bromide*

(a) Gel transfer in an RNase free plastic container and incubate the gel in 0.5 M ammonium acetate containing 0.5µg/mL at room temperature for 40 min.

(b) Carefully transfer the gel to a UV transilluminator and takes photographs using a digital gel documentation system.

5. *Results*

The final result of total RNA run on a denaturing gel represents sharp clear visual 28S and 18SrRNA bands. 28S rRNA band expressed twice as extreme as compared to the 18S rRNA band (**Figure 16.**, lane 3). When the ratio of 28S:18S is (2:1 ratio) that represent RNA is completely intact. In case partly degraded RNA shows a smeared expression that have lack the sharp rRNA bands and not exhibit the 2:1 ratio of a good high-quality RNA. Moreover, completely degraded RNA represents low molecular weight smear (**Figure 16.2**, lane 2). Markers run on the gel allow the proper size of every band to be determined and represent a good control that ensure the gel was run properly (**Figure 16.1,** lane 1). Basically, 200 ng of RNA must be loaded onto a denaturing agarose gel and easily visualised with EtBr.

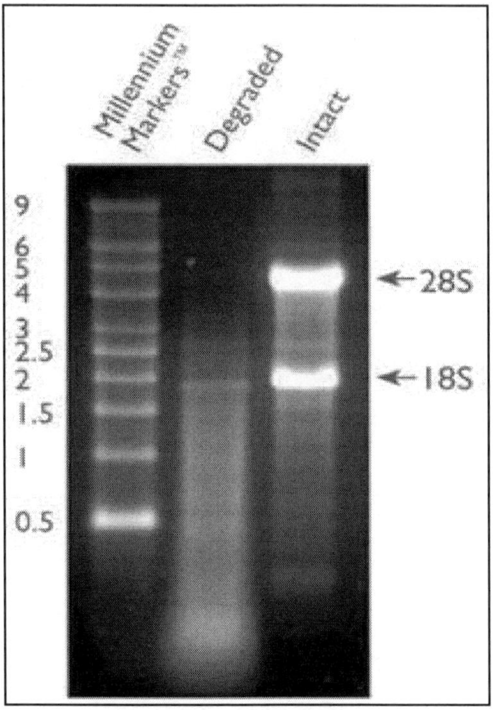

Figure 16.1: The difference between Intact and Degraded RNA.

2. RNA Quantification using UV Absorbance

Introduction

Ultraviolet (UV) absorbance is well-established method used to measure biological substances.

Among these substances, nucleic acid (DNA and RNA) and protein is most routinely measured in molecular biology laboratories. For calculating the concentration and purity of nucleic acid and protein samples is potentially important for various downstream process like DNA microarrays, PCR, qPCR and sequencing.

Spectrophotometer is capable to determine the concentration of DNA and RNA present in the solution, as well as their purity. The spectrophotometer analysis basically worked on the principles that nucleic acid (DNA and RNA) absorbs ultraviolet (UV) light in a specific pattern. The nucleic acid sample is exposed to ultraviolet light at a 260 nm wavelength. Photo-detector assessed the light that passes through the sample. Sample absorbed maximum light, represent the higher concentration of DNA and RNA in the sample. Mainly three most common wavelengths such as 260 nm, 280 nm and 230 nm used for quantification of nucleic acid. For measurement of RNA in the solution, absorbance at 260 nm, because at this wavelength maximum absorption of RNA. Concentration of RNA in solution can be assessed by using the 260 nm reading and a conversion factor based on the extinction coefficient (A260 of 1.0 = 40μg/mL for RNA). In case the amino acids absorb at 280 nm wavelength, at this wavelength generally used to estimate the concentration of protein in the sample. Moreover, take absorbance at 230 nm wavelength are used to determine the amounts of other impurities like guanidine thiocyanate present in the sample (**Figure 16.2**).

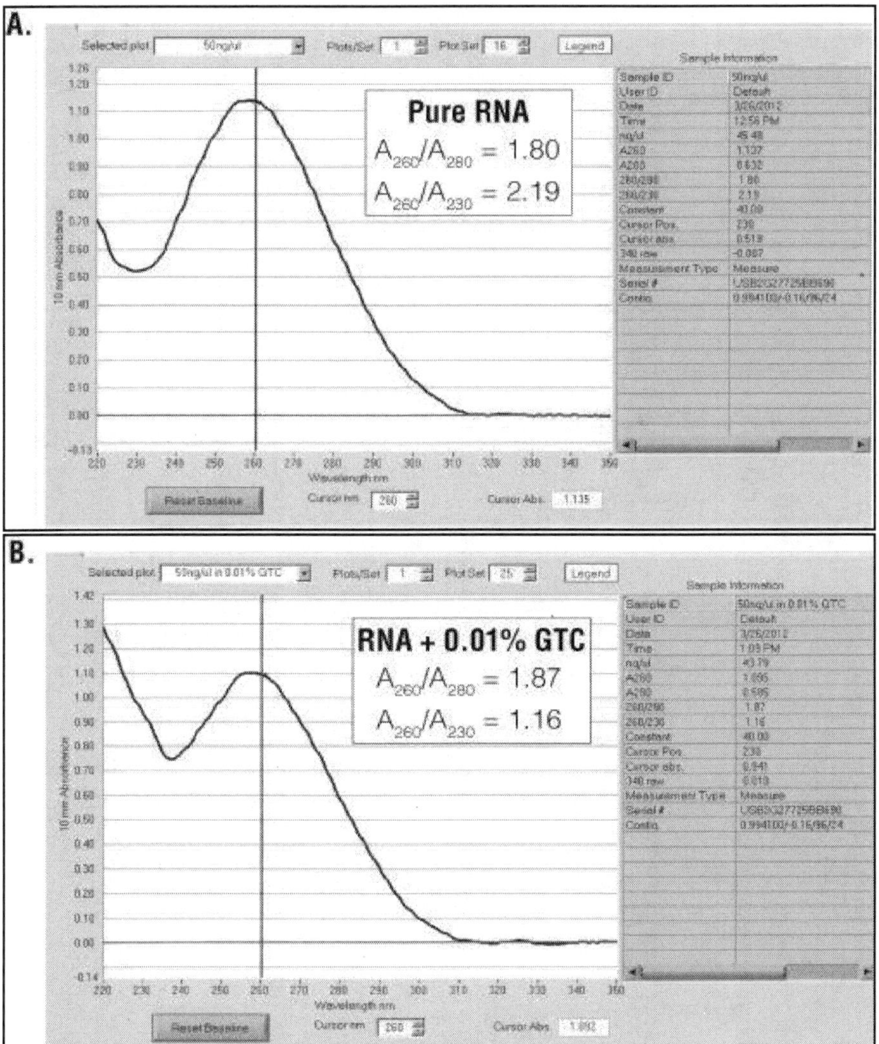

Figure 16.2: NanoDrop® Spectrophotometer Results of RNA with other Contaminants. (A) Pure RNA sample, (B) RNA sample with 0.1 per cent guanidine thiocyanate containing buffer, results indicate reduced purity as determined by the A260/A230 ratio.

Sample Purity Analysis

To determine the purity of nucleic acids, the ratio of the absorbance of nucleic acids and absorbance of the contaminants is calculated. Mostly, requirements for A260/A280 ratios are 1.8-2.2 and requirements for A260/A230 ratios is approximately 1.7.

Demerits of Spectrophotometer and Used NanoDrop® Instrument

Generally, spectrophotometers require cuvettes to analysis the sample and require more than mL of sample volume. To solve this problem, used instruments such as the NanoDrop® (Thermo Scientific) spectrophotometers, it requires small amount of sample volumes for measuring the observance. The powerful NanoDrop® instrument uses 190-840nm wavelength to measure absorbance and it requires small volume of sample range from 0.5-2µl for an accurate determination. For RNA estimation, this instrument detects a minimum of 2ng/µl up to 12,000ng/µl. NanoDrop® instrument have wide range of detection, require small amount of sample, samples mainly do not need to be diluted before measurement, and a single measurement can be taken in less than 30 seconds. Unlike other quantification methods, this method no required other chemical reagents for analysis.

The Beer-Lambert Law is mainly related the amount of light absorbed to the concentration of the absorbing materials. This law states that there is a logarithmic dependence between the transmission of light through a substance and the product of the absorption coefficient of the substance and the path length. UV light is passed through the sample at a specified path length, and concentrations of nucleic acids can be directly calculated by measuring absorbance values at 260 nm against a blank using the Beer-Lambert's equation:

$$A = \varepsilon \times c \times l$$

where,

A = UV absorbance at a particular wavelength ε = extinction coefficient

c = Concentration of sample

l = Path length of a spectrophotometer

This law can be used extinction coefficient to examine the concentration of nucleic acid. **Concentration (ng/mL) = [Absorbance × extinction coefficient]/Path length (cm)**

Extinction coefficients:

dsDNA:50 ng-cm/µl

ssDNA: 33 ng-cm/µl

RNA: 40 ng-cm/µl

Nanodrop Analysis

Procedure

1. First, wash the nanodrop by pure distilled water and then 2µl pure water add in nanodrop as a blank.

2. Change the all-computer settings to analysis the RNA sample and select blank.
3. After settings, add 2µl sample and select measure.
4. Record the amount of RNA in the sample (ng/µl).
5. Check the A260/A280 and A230/A260 ratio values.

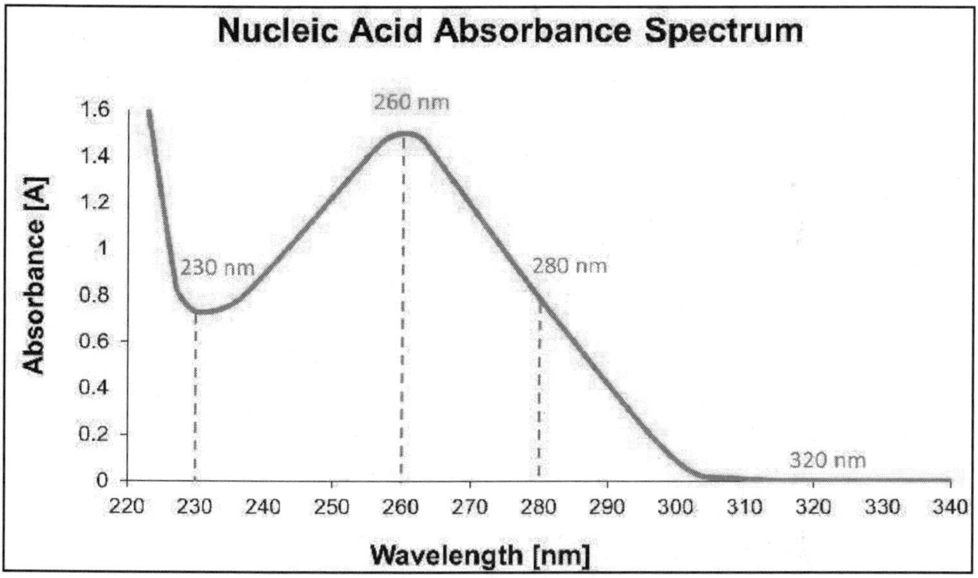

Figure 16.3: Absorbance Spectrum of Nucleic Acids with Relevant Wavelengths

3. Real time PCR (qPCR)

Introduction

Real time PCR also known as quantitative PCR (qPCR) is most popular technique of nucleic acid (DNA and RNA) quantification. This technique basically worked on thermal cycling comprising of repeated cycles of heating and cooling for DNA melting and enzymatic replication of targeted amplicons by polymerase enzyme. In qPCR method, total amount of amplified product can be determined at the end of each cycle amplification or in real time during the exponential phase of amplification. After completion of every cycle of amplification, accumulates amplified product and reaches a threshold level, where it can be easily detected. Some fluorescence dye can be used to measured amount of amplified product such as SYBR® Green and labeled DNA probe (Taqman®).

RT-qPCR is very sensitive and used to determine the amount of a particular RNA target by using reverse transcription (RT) step followed by qPCR. Primer designing the most process for making successful RT-qPCR. For RNA (cDNA) detection, specific RNA primers are used that flank an intron of the target sequence. In control reaction, no reverse transcriptase is added to the RT process.

Threshold is the level of fluorescence, at which product accumulation examined from the background. This threshold level is automatically set above the baseline, within the exponential place. The cycle number represent the amplification product crosses that detection threshold is the Quantification cycle (Cq) **(Figure 16.4)**.

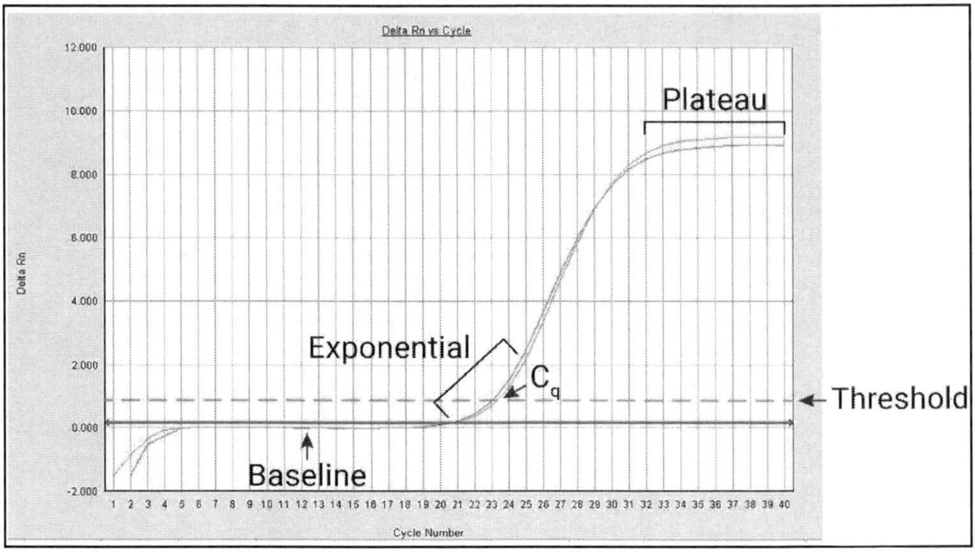

Figure 16.4: Real Time PCR Amplification Plot.

When unknown sample concentration can be measured by corresponding Cq number compared to a standard curve of Cq values (**Figure 16.5**).

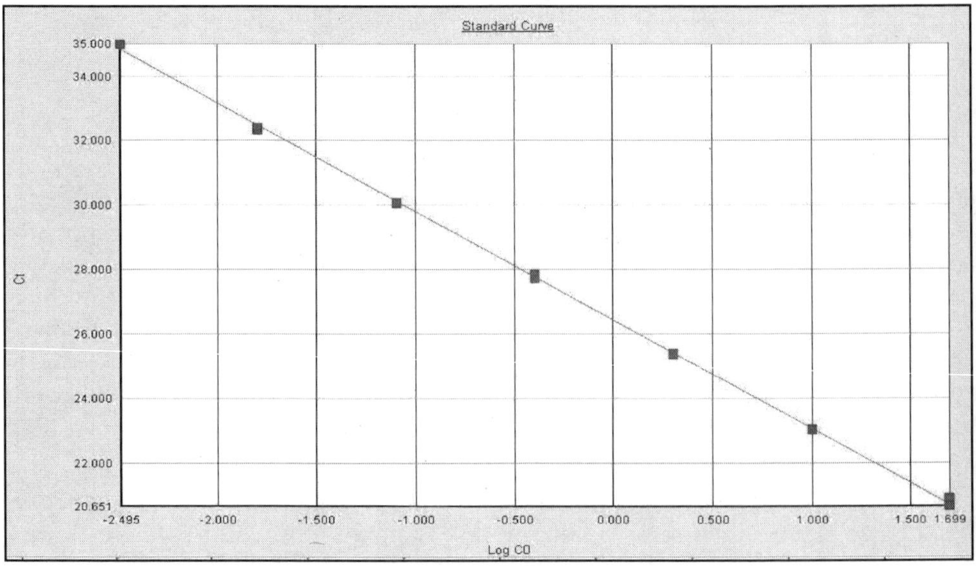

Figure 15.5: Standard Curve of a Real Time PCR.

Compression of RNA Analysis Methods

RNA Analysis Method Characteristics	*UV Absorbance*	*Dye Based Quantification*	*Agarose Gel*	*2100 Bioanalyzer*	*RT-qPCR*
Integrity	Not determine	Not determine	Determine	Determine	Delayed Cq value indicate poor integrity
Concentration	Determine	Determine	Qualitative andquantitative	Determine	Determine
Purity	Determine	Not determine	Not determine	Not determine	Delayed Cq value indicate low purity
Specificity- DNA/ RNA	Not determine	Depends on dye and system used	Use nucleic acid specific dyes	Not determine	Determine
Equipment use	Spectrophotometer	Fluorometer	Gel box,Stain	Bioanalyzer	Real time instrument
Assay time	Less than 1 min	15-30 min	10-120 min	1 h	1-2 h
Liner range of detection	NanoDrop Spectrophotometer: 2-12,000 ng/µl	RNA assay Kit: 0.2-200 ng	Ethidium bromide: less than 34ng SYBR Green: less than 14ng	Agilent RNA 600 Nano Kit: 5-500ng/µl	Mainly depends on RNA quality and primer design

General Steps for Real Time PCR

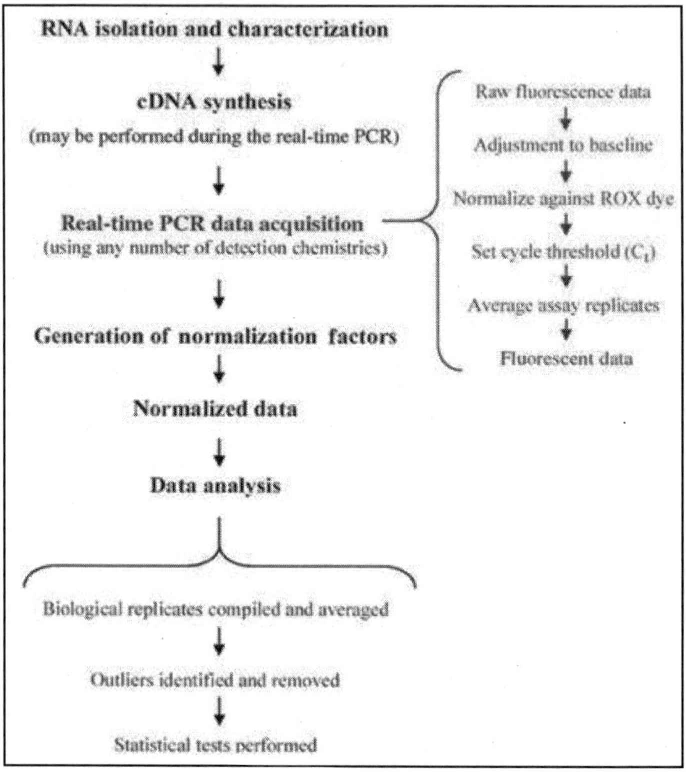

Figure 16.6: Steps of Real Time PCR Used for Measuring Gene Expression.

Chapter 17

Agarose Gel Electrophoresis (Armstrong *et al.*, 2015)

What is Electrophoresis

Electrophoresis mostly used for separation and purification of biomolecules namely proteins and nucleic acids (DNA and RNA) based on their net charge, size and conformation on a matrix. Nucleic acid is negatively charged macromolecules due to the presence of phosphate. So that, they move towards anode side at the migration rate and mainly depends on their size. Basically, proteins molecules have positive or negative charge; this make enables the movement of a protein molecule towards a pH called isoelectric point at which the molecule not carried any net charge. By denaturing the proteins and give a uniform charge, it is sure that the separation based on the size. The biomolecules are electrophoresed within a gel and gel made up of agarose or polyacrylamide. The matrix or gels made up of such types of polymers comprised pores through which the nucleic acid (DNA and RNA) and proteins can pass when voltage is applied.

Agarose Gel Electrophoresis for DNA

Agarose gel electrophoresis is a basic technique in molecular biology and highly effective method for separating, identifying and purifying DNA fragment. Agarose is a polysaccharide extracted from seaweed genera *Gelidium* and *Gracilaria* and consists agarobiose (L-and D- galactose) subunits, used at 0.5 to 2 per cent concentration. Presently standard agarose gels separate DNA fragments ranges from 0.5 to 25 kb, but in case of pulsed-field agarose gels separate DNA molecules from 10 to 2000 kb. Electrophoresis basically used to analyze DNA fragments generated by restriction enzymes and also used for determining the plasmid size and their concentration. Because in the presence of many phosphates in DNA structure, it carries negative charged. With this reference, we can put our DNA samples in agarose gel and use an electrical platform to separate various sizes of DNAs. In agarose gel small size of DNA fragments run faster, but larger size of DNA fragments run in little distance at the same time.

Materials Required

- ☆ Electrophoresis chamber
- ☆ Power supply
- ☆ Casting trays
- ☆ Combs
- ☆ Microwave
- ☆ UV light

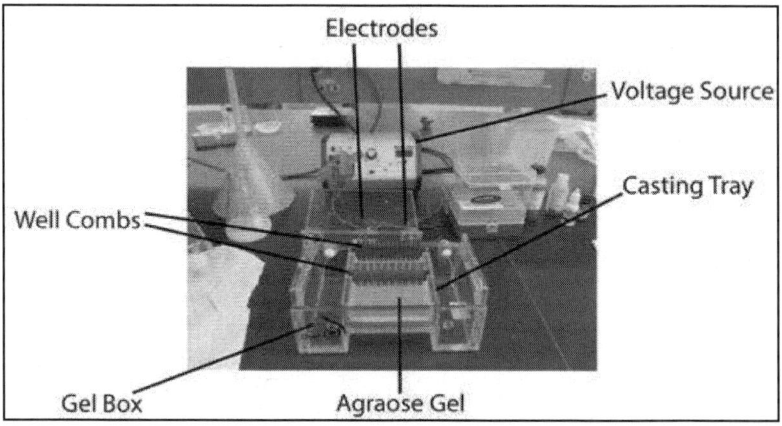

Figure 17.1: components of Gel Electrophoresis.

Electrophoresis Running Buffer

The two most common buffers used.

1. ***50 X TAE buffer (pH 8.0)***

   Tris base: 242 g (0.89 M)

   Glacial acetic acid: 57.1 mL (0.89

   EDTA 0.5 M: 100 mL (0.02 M)

   ddH_2O- 1 L

2. ***10X TBE buffer (pH 8.0)***

   Tris Cl: 108 g (0.89 M)

   Boric acid: 55 g (0.89 M)

   EDTA 0.5 M: 40 mL (0.02M)

   ddH_2O- 1 L

Loading Dye

This is mixed with DNA samples for use in agarose gel electrophoresis. It generally contains a dye to assess how "fast" your gel is running and a reagent to render your samples denser than the running buffer.

☆ Ethidium bromide (0.5 mg/100 mL distilled water)

Sl.No.	*Dye*	*0.1 to 1.5 per cent Agarose*	*2.0 to 3.0 per cent Agarose*
1.	Xylene cyanol	10,000 - 4000bp	750-200 bp
2.	Cresol red	2000-1000 bp	200-150 bp
3.	Bromophenolblue	500-400 bp	150-50 bp
4.	Orange G	Less than 100 bp	-

Prepare of Agarose Gel Solution

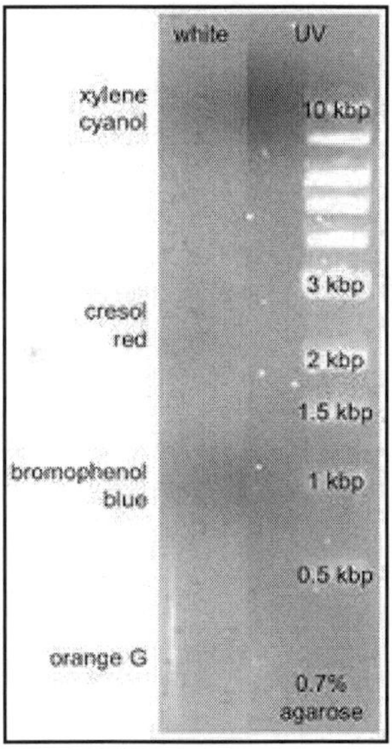

1. Take 1 gm of agarose and dissolve in 100 mL of TAE/TBE buffer in a beaker or flask (mostly we used 0.5, 1 and 1.5 per cent gel) and then heated in a microwave oven to make the agarose completely dissolved in buffer.

   Note: The amount of agarose mainly depends on the percentage of the gel required. The volume of agarose solution must be prepared according to the size of the gel tray used.
2. Allow to agarose solution cool down to about 50°C for 5 min and add ethidium bromide (0.5 μg/mL final concentration).

   Note: Ethidium bromide is a carcinogenic chemical. Used very carefully and wear gloves and masks.
3. After cooling of agarose, prepare the gel tray for casting the gel. Placed the comb in the grooves.
4. Pour agarose solution in gel tray and ensure that equally distributed on the tray with no air bubbles trapped in the gel.
5. Put in room temperature for 25-30 mins, until it has completely solidified.

Appropriate Agarose Concentrations for Separating DNA Fragments of Various Sizes

Sl.No.	*Agarose (g/100mL)*	*Effective Range of Resolution of Linear DNA Fragments (kb)*
1.	0.5	1-30
2.	0.7	0.8-12

Sl.No.	Agarose (g/100mL)	Effective Range of Resolution of Linear DNA Fragments (kb)
3.	1.0	0.5-10
4.	1.2	0.4-7
5.	1.5	0.2-3

Precast Agarose Gel

Sigma-Aldrich offers precast agarose gels in 8-, 20- and 24-well format with added ethidium bromide. Continue to "Running the gel" step if precast gels are being used.

Product No.	Agarose content	Well Format	Suitable for	Orientation
P5472	1 per cent	8-well	DNA	Portrait
P5722	1 per cent	20-well	DNA	Landscape
P5972	1 per cent	24-well	DNA	Portrait
P6097	4 per cent	24-well	DNA	Portrait

Preparation of Samples

1. Isolate DNA from plant tissue with the help of standard protocol.
2. Required lower amount of DNA sample for better detection on agarose gel.
3. Mix the samples with 10X loading buffer. Mostly we used 3 µl loading buffer.

Sample Loading and Running the Gel

1. After solidification of gel, remove the combs and transfer the gel with the tray to the main tank.

   Note: Make sure that the wells of your gel are at the black Ru (negative) end of the gel box. DNA has a negative charge, and runs toward a positive charge (the red side).
2. Pour buffer into the gel box until the gel is fully covered with buffer and all wells are filled with buffer.
3. Pipet 1 µg of sample onto Parafilm and add distilled water to bring drop volume to approximately 10 µl.
4. Add 1 µl loading dye very carefully to each 10 µl DNA sample.
5. Add 7 µl ladder carefully by pipette into the first lane of the gel.
6. Load sample in each separated well by pipetting.

7. Cover the lid on the tank and connect power supply and run 90-110 voltage for 45 mins.
8. Until the blue dye has run about 2/3 across the gel.

Gel Staining

1. After completion of electrophoresis, transfer the gel to UV light and acquire the photos of gel.
2. The samples will appear as bright bands.

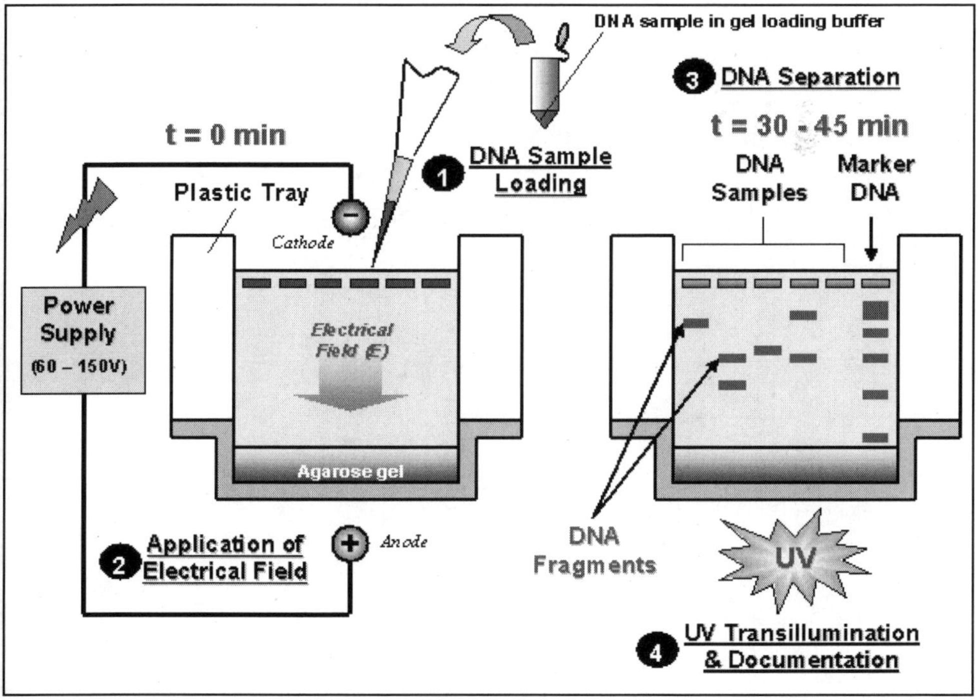

Figure 17.2: Procedure of Gel Electrophoresis.

Chapter 18

Polymerase Chain Reaction (PCR) Technique

Introduction

The development of PCR has been the most valuable invention that revolutionary changed in the field of molecular science. The procedure of PCR was described and experimentally proved by Kary Mullis in 1985. The revolution and refinement of this technique with the potential introduction of a thermal stable DNA polymerase extracted from *Thermus aquaticus* bacterium, and called *Taq* DNA polymerase.

PCR is most potential amplification technique that can produce billions copies of DNA (amplicon) in few hours from small amount of DNA templates or target sequence. For instance, small quantity of DNA that is, nano gram (ng) can be amplified to achieve micro gram (µg) of DNA by application of this technique. Basically, PCR technique works based on the in vitro enzymatic amplification of a particular target sequence of DNA in a cyclic process by using two oligonucleotides. These two oligonucleotides basically a primer and those primers flank the DNA fragments to be amplified and employs repeated cycles of heat denaturation of the DNA, annealing of the primers to their complementary sequences, and extension of the annealed primers with a thermostable DNA polymerase. These oligos primers have different sequences and are complementary to the sequences on the opposite strands of the template DNA and flank the segment of target DNA that is to be amplified. Extension products themselves are also complementary to and capable of binding primers, successive cycles of amplification essentially double the amount of target DNA synthesized in the previous cycle.

Principle of PCR

PCR uses the enzyme DNA polymerase that directs the synthesis of DNA from deoxynucleotide substrates on a single-stranded DNA template. DNA polymerase adds nucleotides to the 3′ end of a custom-designed oligonucleotide when it is annealed to a longer template DNA. Thus, if a synthetic oligonucleotide is annealed to a single-stranded template that contains a region complementary to the oligonucleotide, DNA polymerase can use the oligonucleotide as a primer and elongate its 3′ end to generate an extended region of double stranded DNA.

Two Primers as Initial Foundation in PCR Technique

PCR utilizes the DNA polymerase enzyme, which naturally catalyzes the synthesis of DNA sequences. All DNA polymerases synthesize DNA in the 5′ to 3′ direction. In order to start the synthesis process, it is necessary to have a pair of chemically synthesized oligonucleotide primers made of DNA (as the primers are synthesized by cells).

Components of PCR

1. DNA template
2. DNA polymerase enzyme
3. Oligonucleotide primers (Forward and reverse)
4. Deoxyribonucleotide triphosphate
5. Buffer system
6. PCR tube

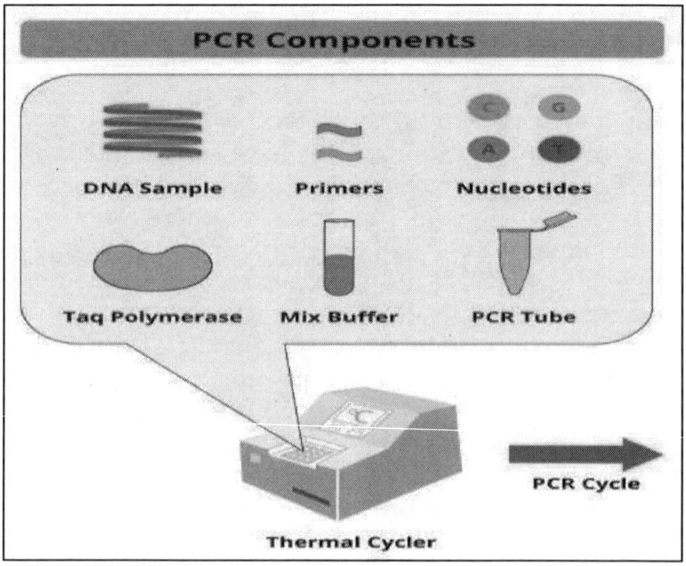

PCR Reaction Mixture

Sl.No.	Reagents	Quantity (μl)
1.	Taq buffer (10X)	2.5
2.	dNTPs (2.5 mM)	0.5
3.	Forward primer (10 pmols/μl)	1.0
4.	Reverse primer (10 pmols/μl)	1.0
5.	Template DNA (25 ng/μl)	2.0
6.	Taq polymerase (5 U/μl)	0.1
7.	Millipore sterilized water	17.9
	Total	**25.0**

Steps of PCR

1. *Denaturation*

Denaturation occurs when the reaction mixture is heated to 94°C for about 0.5 to 2 minutes. This breaks the hydrogen bonds between the two strands of DNA and converts it into a single-stranded DNA. The single strands now act as a template for the production of new strands of DNA. The temperature should be provided for a longer time to ensure the separation of the two strands.

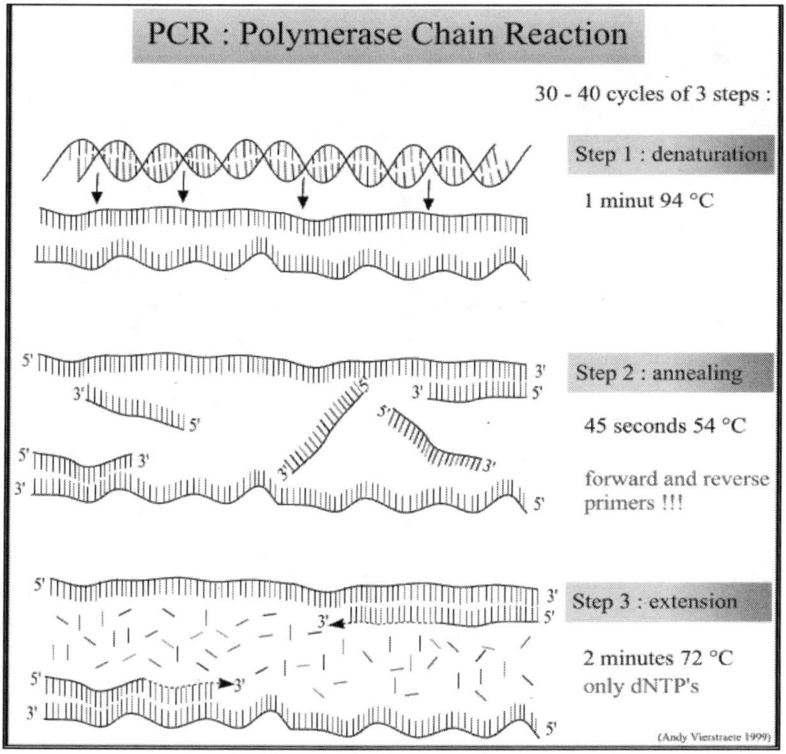

2. *Annealing*

The reaction temperature is lowered to 54°C for around 20-40 seconds. Here, the primers bind to their complementary sequences on the template DNA. Primers are single-strand sequences of DNA or RNA around 20 to 30 bases in length. They serve as the starting point for the synthesis of DNA. The two separated strands run in the opposite direction and consequently there are two primers- a forward primer and a reverse primer.

3. *Extension*

At this step, the temperature is raised to 72°C. The bases are added to the 3′ end of the primer by the Taq polymerase enzyme. This elongates the DNA in the 5′ to 3′ direction. The DNA polymerase adds about 1000bp/minute under optimum conditions. Taq Polymerase can tolerate very high temperatures. It attaches to the primer and adds DNA bases to the single strand. As a result, a double-stranded DNA molecule is obtained. These three steps are repeated 20-40 times in order to obtain a number of sequences of DNA of interest in a very short time period.

Procedure of PCR

1. *Add Reagents to PCR Tubes*

For a large number of reactions, a master mix without the template should be set up and aliquoted into reaction tubes. At the end, template should be added to appropriate tubes.

2. *Mix and Centrifuge*

Note: Add 50 µL of mineral oil to the top of each tube to prevent evaporation if using a thermal cycler without a heated lid.

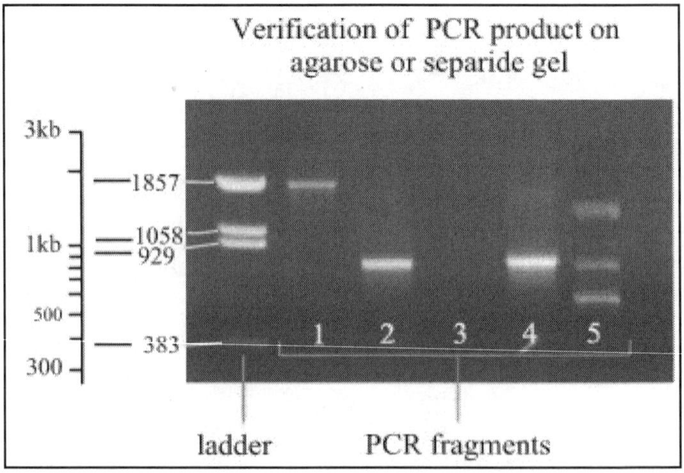

Figure 18.1: Verification of the PCR Product on Gel.

3. Amplify

The amplification parameters will vary depending on the primers and the thermal cycler used. It may be necessary to optimize the system for individual primers, template, and thermal cycler.

4. PCR Product Detection and Analysis

The product of a PCR comprised of one or more DNA fragments. The detection and analysis can be done by agarose gel electrophoresis. The DNA is revealed by ethidium bromide staining.

Note

Mg^{++} and additives: Mg^{++} concentration of 1.5–2.0 mM is optimal for most PCR products generated with *Taq* DNA Polymerase. The final Mg^{++} concentration in 1X Standard *Taq* Reaction Buffer is 1.5 mM. This supports satisfactory amplification of most amplicons. However, Mg^{++} can be further optimized in 0.5- or 1.0-mM increments using $MgCl_2$.

Amplification of some difficult targets, like GC-rich sequences, may be improved with additives, such as DMSO and formamide.

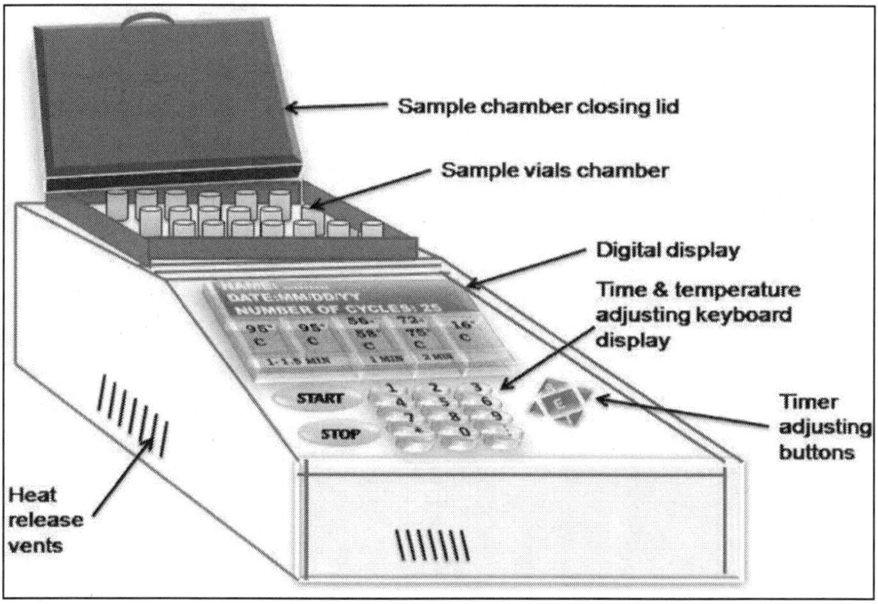

Figure 18.2: Set Up of PCR Machine.

Chapter-19

High Performance Liquid Chromatography (HPLC)

Introduction

In the 1970s, the term High performance liquid chromatography (HPLC) was introduced to characterize the modern HPLC technique belonging from the classical low-pressure column chromatography. HPLC is a robust and most powerful separation technique used for isolation of wide ranges natural products and also analysis of organic and inorganic biomolecules and proteins. Mainly, HPLC is a chromatographic method that can applied for separation of mixture compounds and used in various field such as phytochemical, biochemistry, environmental science and plant science to identify, quantify and purity the specific molecules of the mixture. This technique classically based on the mechanisms of adsorption, partition, and ion exchange, which mainly depend on the various type stationary phase. HPLC have solid stationary phase, which enclosed in a stainless-steel column, and a liquid mobile phase. The required molecules can be separated based on the difference in the relative distribution ratio of the solutes between the two phases.

Principle

The high-performance liquid chromatography works on the principle that some components of the sample take longer than others to pass through the chromatography column. The time taken by the molecules depends on the affinity of the molecule with the mobile phase (liquid or gas) and the stationary phase (solid or liquid). The ones with greater affinity with the stationary phase take longer to pass through the column and vice-versa.

Materials Required

1. Chromatographic column

2. Pumping system: The pump is used to ensure an adjustable, constant, and reproducible flow of 1.0 mL/minto the analytical column.
3. Injection value: The injector introduces a narrow stream of variable and reproducible sample volume onto the top of the analytical column.
4. Stationary and mobile phase
5. Connecting tube
6. Data collecting device: The data collecting device acquires, stores data, and perform calculations.
7. Detector: The detector provides the system with high sensitivity, linearity, and long-term stability to detect the signals from the chromatographic column. UV detectors are most popular among all the detectors because they offer high sensitivity and also because majority of naturally occurring compounds encountered have some UV absorbance.

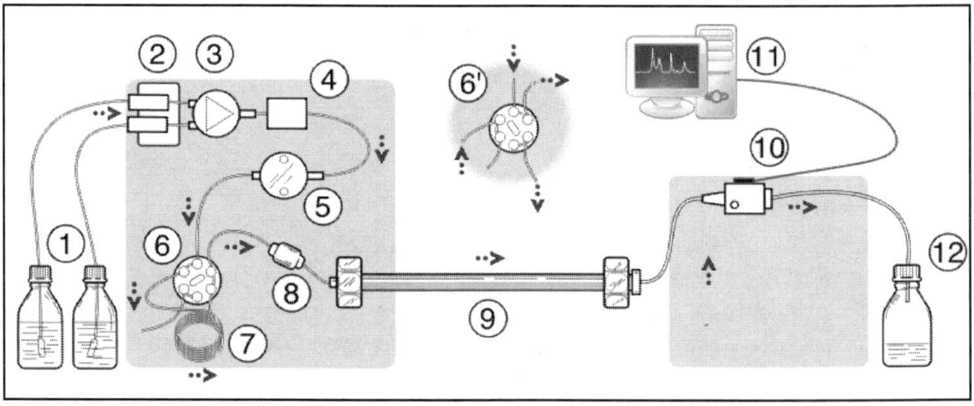

Figure 19.1: Schematic Diagram of HPLC.

(1) Solvent reservoirs, (2) Solvent degasser, (3) Gradient value, (4) Mixing vessel for delivery of the mobile phase, (5) High- pressure pump, (6) Switching value of inject position, (7) Sample injection loop, (8) Pre-column, (9) Analytical column, (10) Detector, (11) Data acquisition, (12) Waste collector.

Checking the Settings of HPLC System

1. Ensure that the waste line is in the waste container.
2. Set the flow rate of the mobile phase to 0.5 mL/min.
3. Set the minimum and maximum pressure and the flow rate to the correct values on the front panel of the delivery pump.
4. Set the minimum pressure at 250 psi.
5. Set the maximum pressure at 4,000 psi.
6. Press "zero" on the detector's front panel to set at the pure mobile phase.
7. Rinse a 100-μL syringe with deionized water, then wash it with several volumes of component solutions to be analyzed, and fill the syringe.

Sample Preparation for HPLC

1. Take 1 gm of fresh leaf sample collected from each treatment and prepare extract by homogenized in 10 mL of 80 per cent methanol containing 5 N HCl in a mortar and pestle.
2. Keep the extract overnight at room temperature and then centrifuge at 13,000 rpm for 10 min and filter by the help of sterilized Whatman No. 1 filter paper.
3. Fractionate the filtrate again three times by using ethyl acetate.
4. Collect the ethyl acetate fractions and evaporate it under reduced pressure on rota evaporator (Eyela N-N series, Tokyo, Japan).
5. Resuspend the dried residue in 1 mL HPLC grade methanol and filter by using a membrane filter (0.2 µm).

HPLC Analysis

1. HPLC system Shimadzu model LC-20 AD, Japan was equipped with two LC-20 AD pumps system and a detector SPD-10 A used.
2. The filtered 20 µl sample is injected stationary phase consisted of C-18 HPLC column (4 µm, 250×4.6 mm, Phenomenex, USA).
3. Similarly, separation of compounds is achieved with mobile phase comprised of a two-pump linear gradient solvent system for pump A (acetonitrile 100 per cent HPLC grade) and pump B (1 per cent acetic acid in HPLC grade) with a flow rate of 1.0 mL min^{-1} for 30 min.
4. The program is initially started with acetonitrile from pump B, changing to 32 per cent in 15 and finally to 50 per cent in 30 min which is following by 25 min washing.
5. Identification of molecules of samples by comprising retention time (Rt) with those from authentic standard at 250 nm.
6. The peak of the samples (µg/g FW) and data is integrated by class VP series software, Shimadzu, Japan.

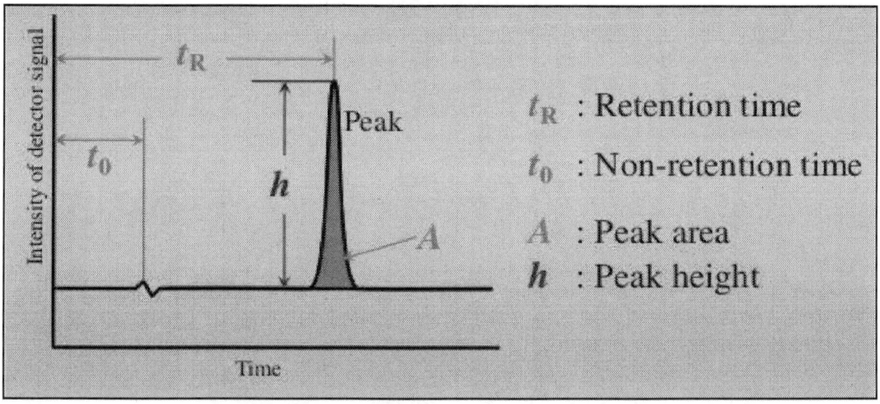

Figure 19.2: HPLC Chromatogram.

Calculation

1. Calculate the concentration of all the components in the sample solution.
2. Determine the peak area on the chromatograms for each standard and the unknown samples by the triangular method, which equals peak height times the width at ½ height.
3. Determine the time taken by the components to reach the peaks.
4. Draw calibration curves of peak area vs. concentration (mg/L) for all components.
5. Calculate the least-squares fit for calibration curves.

Notes

1. To clean the system and equilibrate it, you need to run enough solvent.
2. Open the instrument and observe the instrument-related data first, especially the column pressure. Low column pressure means there have occurred some errors. Reset's column pressure limits.
3. The sample must be filtered before injected, otherwise the column may be clogged.
4. Add mobile phase (water, methanol or acetonitrile) in time to avoid the lack of mobile phase caused by instrument failure.
5. Before injection, use a sample solution or mobile phase to rinse the syringe to avoid errors.
6. Before loading, push the liquid to the position of the needle head to prevent air bubbles from entering.
7. Always use HPLC grade water and reliable solvents.
8. Wash the column and pump with water to prevent deposition of solid crystals which could lead to column blockages and faster wear of pump components.

Chapter 20

Primer Designing for PCR

Introduction

In the last decades, polymerase chain reaction plays a vital role in molecular biology and PCR- based technique have been successfully applied for genetic identification of many taxa of various disease-causing plant pathogens. Small amount of genetic material can be amplified and get huge number of copies by using PCR technique. A successful PCR technique comprised primers and target sequence, and both can affect the efficiency, accuracy of PCR assays. So, care must be taken when choosing a good primer. In PCR technique primer designing play pivotal role for better performance.

Primer is a small synthetic oligonucleotide which may be complementary to target sites on the template DNA in a PCR. Literally, "to prime" means to "initiate" or "start". *In vivo*, a short oligo sequence (*i.e.*, the primer) is required, because the DNA polymerase enzyme has no capacity to start the DNA replication without adding primer. Primers also contain guanine or cytosine, or both at the 3′-ends to enhance the capacity of oligonucleotide binding and form a stable duplex with target DNA at the annealing temperature. The primers are the key to the success or failure of a PCR experiment. If the primers are design correctly. The results of experiment in amplification of a single DNA fragment corresponding to the target region of a template sequence.

While designing a good primer, these steps should be followed:

- ☆ Check the databases of present primers
- ☆ Choose a target sequence
- ☆ Design primer
- ☆ Check the primer properties
- ☆ Determine PCR product properties
- ☆ Optimize the protocol

Properties of PCR Primers

1. **Length of the primer:** Primes are short sequences, basically, 18 to 30 bases give best result in PCR amplification. Shorter primer less than 18 bp not good for amplify in each cycle. In case primer short, lower annealing temperature and suppress the capacity of amplification. Moreover, long sequence primer not good because the annealing temperature of long primer is too high.
2. **Primer sequence:** Especially the 3′-end of the primer molecule is critical for the specificity and sensitivity of PCR. Primer pairs should be checked for complementarity at the 3′-end. This often leads to primer-dimer formation. Bases at the 5′-end of the primer are less critical for primer annealing. Therefore, it is possible to add sequence elements, like restriction sites, to the 5′-end of the primer molecule.
3. **Melting temperature (Tm):** The temperature at which half of the primer has annealed to the template. Usually, good results are obtained when the Tm's for both primers are similar (within 2-4 °C) and above 60°C. The Tm for a primer can be estimated using the following formula:

   $Tm = 2^{\circ}C^{*} (A+T) + 4^{\circ}C^{*} (C+G)$
4. **GC content of the primers:** GC content between 40 per cent to 60 per cent is acceptable. The annealing temperature of the primer between 55°C to 65°C with 50 per cent GC performs better. If GC content is too high, at given annealing temperature primer will mismatch with other sequences.
5. **Complementation in forward and reverse primers:** While designing primer, keep in mind that both forward and reverse primer do not match with each other or are not complementary with each other. Otherwise, instead of binding with target sequences, both primers will bind with each other and creates a dimer. Dimers can easily be amplified in PCR because it is shorter sequences of up to 50 to 60 bp.
6. **Repeat bases in primers:** Repeated bases can bind within the primer and makes primer non-active. If some of the dinucleotides or trinucleotide are repeated in primer, it binds within the primer and creates a hairpin loop.
7. **Primer annealing temperature (Ta):** The milting temperature of the primer is the calculated of the DNA-DNA hybrid stability and mostly used to estimating the annealing temperature. When annealing temperature is too high it produces insufficient primer-template hybridization and get low PCR product. But in case annealing temperature is too low it creates non-specific products incited by maximum number of base pair mismatches.

   $Ta = 0.3 \times Tm\ (primer) + 0.7 \times Tm\ (product) - 14.9$
8. **Avoid cross homology:** To improve the capacity, specificity of primer it is important to remove the cross homology. To solve the problem of cross homology by primer designing. When work on BLAST the templates against the appropriate non-redundant database and software will decrease the proper results. You can identify the regions significant

cross homologies in every template and avoid them during searching the primer.

Procedure for Primer Designing

1. Firstly, type the URL, (www.ncbi.nlm.nih.gov/) to open the NCBI (National Centre for Biotechnology Information) home page (**Figure 20.1).**

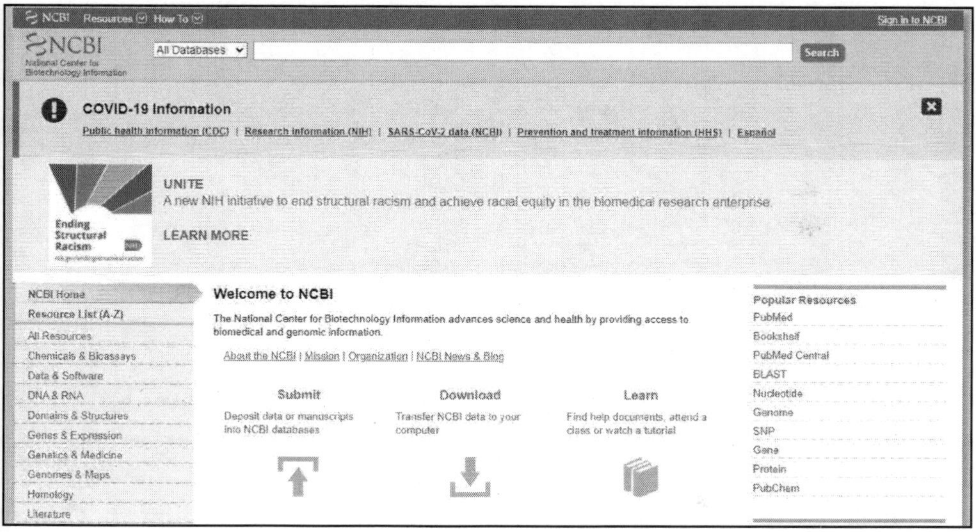

Figure 20.1: Home Page of NCBI.

2. Search the required nucleotide sequence, select the target sequence from NCBI nucleotides and save it in FASTA format (**Figure 20.2).**

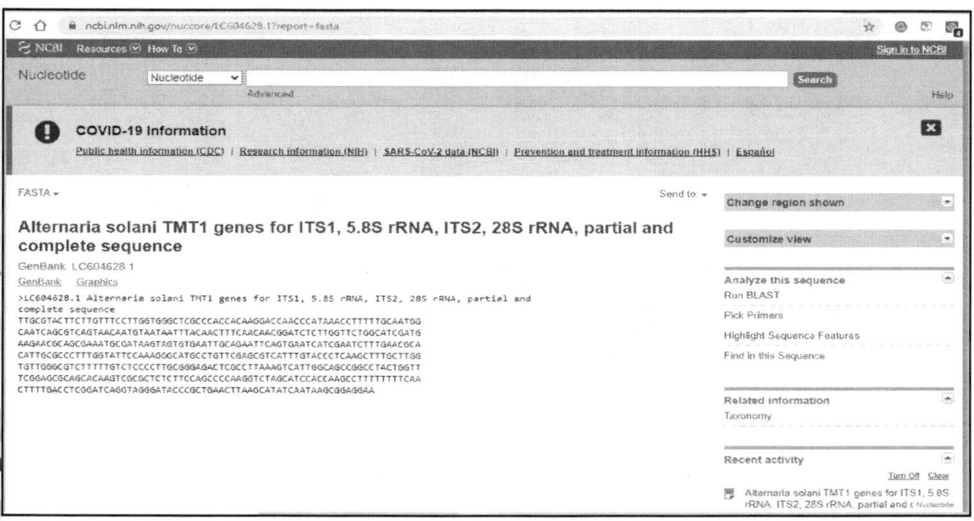

Figure 20.2: FASTA Format.

3. Open the primer 3 (version 0.4.0) software by using the URL (https://bioinfo.ut.ee/primer3-0.4.0/).
4. Paste the nucleotide sequence in FASTA format in the box for source sequence in the primer 3 page (**Figure 20.3**).

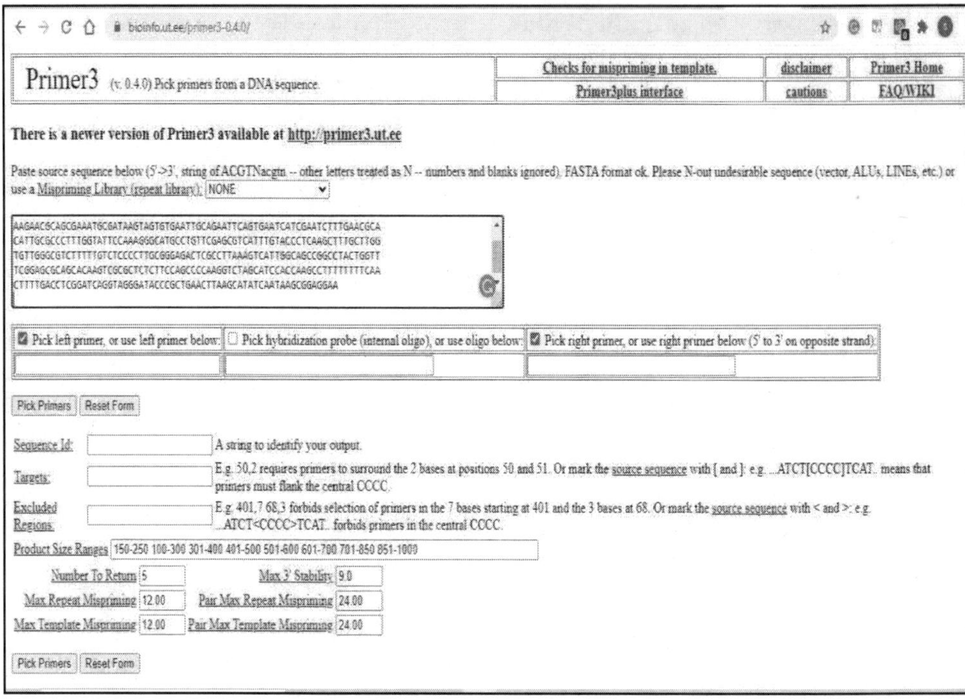

Figure 20.3: Primer 3 Software.

5. All the parameters available on the primer 3 homepage. All parameters not equally important but some parameters important like primer size, melting temperature, GC per cent content, maximum self-complementary and maximum melting temperature difference.
6. When all parameters properly set than click on the "Pick Primers" option to obtain the primers.
7. Primer 3 provides more than one set of primers and there are two set of values starting base of the amplicon is called Forward primer and last base called Reverse primer (**Figure 20.4**).
8. Next step is the primers are to be tested for the checking property by using "Oligo Calc" online software and click on URL (http://biotools.nubic.northwestern.edu/OligoCalc.htmL).
9. Open the Oligo Calc software for oligonucleotides properties checking (**Figure 20.5**).

Figure 20.4: Primer 3 Output of the Primers.

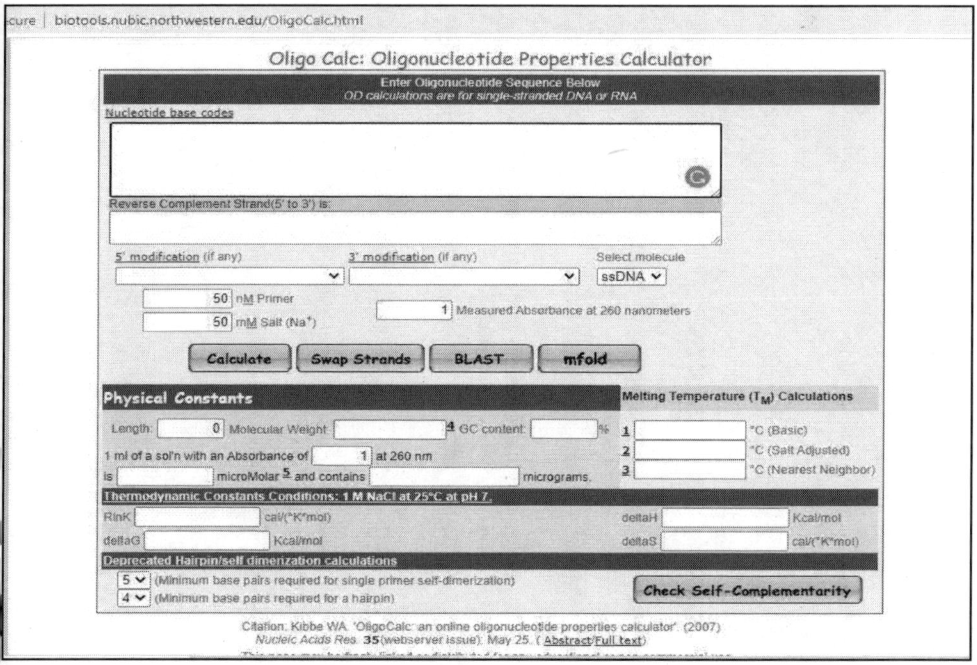

Figure 20.5: Oligo Calc Software.

10. Last step is used BLAST (Basic Local Alignment Search Tool) software (https://blast.ncbi.nlm.nih.gov/Blast.cgi) (**Figure 20.6**).

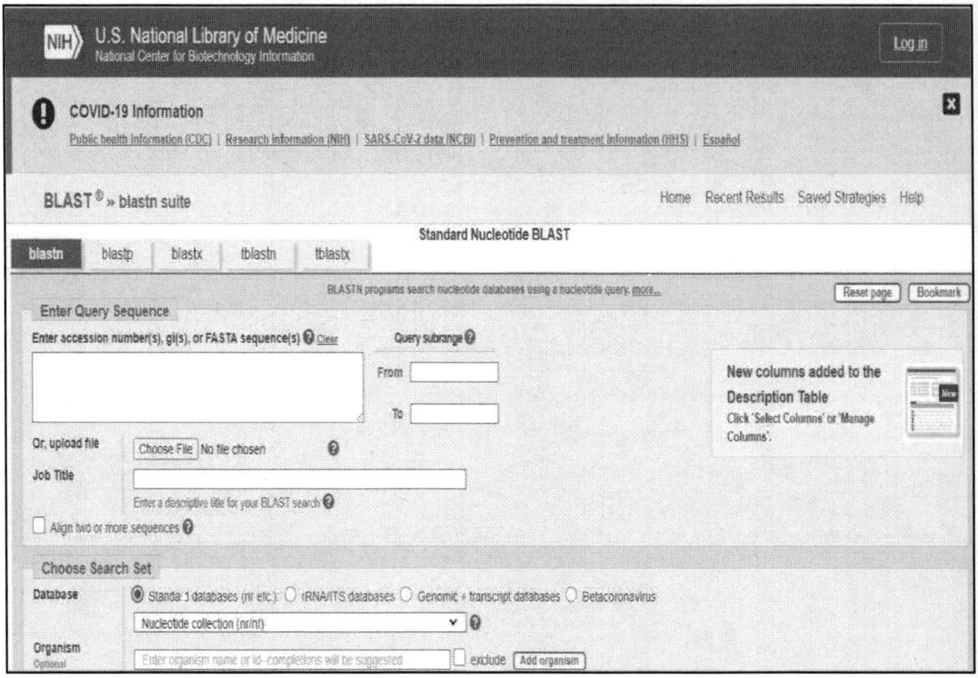

Figure 20.6: BLAST Software.

Chapter 21

Green Fluorescent Protein (GFP) Tagging of the Bacterial Strains and Root Colonization Assay (Shailendra *et al.*, 2020)

Introduction

This technique widely used as excellent reporter molecule in molecular biology stream. The primary structure of green fluorescent protein (GFP) consists of 238 amino acid residues, a secondary structure of hydrogen bonded helices and pleated sheets and tertiary structure is barrel shaped capped with helices. This protein first time isolated from the jelly fish, *Aequorea victoria,* is known as wild-type GFP or wtGFP. Moreover, protein fluoresce green when exposed to blue and UV light. First time, the gene for GFP was successfully inserted into *E. coli* bacteria in 1994. Protein can be tagged with GFP through genetic modification in order for a protein to be visualized under fluorescence microscope.

Green Fluorescent Protein (GFP) Tagging of *Pseudomona geniculate* M 32 and Root Colonization Assay

Materials and Reagents

- ☆ Bacterial strain *Pseudomona geniculate* M 32
- ☆ *E. coli*
- ☆ Confocal Scanning Laser Microscope (Nikon, Japan)
- ☆ Mini-Tn5 gusA::gfp cassette
- ☆ pRK2013 plasmid

- Luria Broth (LB) agar medium
- Primer forward (YL065 (F) 5′-GCGATGTTAATGGGCAAAAA-32)
- Primer reverse (YL066 (R) 5′-TCCATGCCATGTGTAATCCT-32)
- PCR machine
- Agarose gel electrophoresis
- Tomato seeds

Procedure

1. The selected strain *Pseudomona geniculata* M 32 is tagged with Green fluorescent protein (GFP) and visualized under Confocal Scanning Laser Microscope (Nikon, Japan).
2. The mini-Tn5 gusA::gfp cassette is inserted into *Pseudomonas geniculata* (recipient) by tri-parental mating with Tn5 gusA::gfp cassette (pFAJ1820) containing *E. coli* strain (donor) and pRK2013 plasmid harboring *E. coli* HB101 (helper).
3. The trans-conjugants are selected on a half-strength LB agar medium supplemented with kanamycin and nalidixic acid.
4. The trans-conjugant colonies showing fluorescence under confocal microscope are picked up.
5. The presence of GFP in the selected trans-conjugants is also confirmed by PCR amplification using the primers, *viz.* YL065 (F) 5′ GCGATGTTAATGGGCAAAAA- 32 and YL066 (R) 5′-TCCATGCCATGTGTAATCCT-32 .

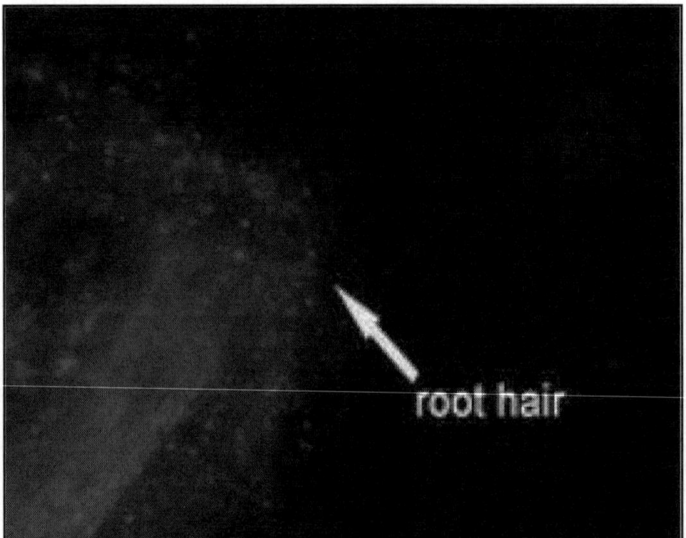

Figure 21.1: Tomato Roots Visualized under Confocal Scanning Laser Microscope Exhibiting Localization of GFP tagged-*P. geniculate* M 32 in the Tomato Roots.

6. The PCR program for the amplification of desired 650-bp amplicon consists of initial denaturation at 94 °C for 3 min, followed by 35 cycles at 94 °C for 30 s, annealing at 56 °C for 1 min, and extension at 72 °C for 1 min with the final extension at 72 °C for 10 min.
7. The amplification is confirmed by agarose gel electrophoresis.
8. The selected GFP-tagged trans-conjugants exhibiting morphological and growth rate characteristics similar to the wild type are used for further experiments.

Root Colonization Assay Procedure

1. The selected GFP-tagged trans-conjugants are used for root colonization assay.
2. The tomato seeds are bio-primed with GFP-tagging *P. geniculate* M 32.
3. Afterward, tomato seeds are sown in the pot containing sterile soil.
4. 15 days after sowing, tomato plants are up-rooted and roots are washed in running tap water.
5. Confocal imaging is done using 488 and 543 nm channels under a confocal scanning laser microscope.

References

1. Ábrahám, E., Hourton-Cabassa, C., Erdei, L., and Szabados, L. (2010). Methods for determination of proline in plants. In *Plant Stress Tolerance,* Humana Press, 317- 331.

2. Airaki, M., *et al.* (2012). Metabolism of reactive oxygen species and reactive nitrogen species in pepper (*Capsicum annuum* L.) plants under low temperature stress. Plant Cell Environ. 35, 281–295.

3. Arjun, S., Kumar, M., Hillol, C., and Pandiyan, K. (2019). National training on Biocontrol strategies for invasive and emerging plant pathogens. ICAR-NBAIM, Kushmaur Mau.

4. Armstrong, J.A. and Schulz, J.R. (2015). Agarose gel electrophoresis. *Curr. Protoc. Essen. Lab. Tech.* 10:7.2.1-7.2.22.

5. Azevedo, R. A. D., Alas, R. M., Smith, R. J., and Lea, P. J. (1998). Response of antioxidant enzymes to transfer from elevated carbon dioxide to air and ozone fumigation, in the leaves and roots of wild type and a catalase deficient mutant of barley. *Physiologia Plantarum*, 104(2), 280-292.

6. Brawerman, G. (1974) In: *Methods in Enzymol.* vol. 30 (Eds. Moldave, K. and Grossman, L.) Academic Press, New York, p 605.

7. Cakmak, I., and Horst, W.J. (1991). Effect of aluminium on lipid peroxidation, superoxide dismutase, catalase and peroxidase activities in root tips of soybean (*Glycine max*), *Physiologia Planatarum*, 83, 463-468.

8. Dhindsa, R.S., Plumb-Dhindsa, P.A.M.E.L.A., and Thorpe, T.A. (1981). Leaf senescence: correlated with increased levels of membrane permeability and lipid peroxidation, and decreased levels of superoxide dismutase and catalase. *Journal of Experimental Botany*, 32(1), 93-101.

9. Dickerson, D.P., Pascholati, S.F., Hagerman, A.E., Butler, L.G., and Nicholson, R.L. (1984). Phenylalanine ammonia-lyase and hydroxycinnamate: CoA ligase in maize mesocotyls inoculated with *Helminthosporium maydis* or *Helminthosporium carbonum*. *Physiological plant pathology*, 25(2), 111-123.

10. Elavarthi, S., and Martin, B. (2010). Spectrophotometric assays for antioxidant enzymes in plants. In *Plant Stress Tolerance,* Humana Press, 273-280.

11. Giri, A.P., Harsulkar, A.M., Patankar, A.G., Gupta, V.S., Sainani, M.N., Deshpande, V.V., and Ranjekar, P.K. (1998). Association of induction of protease and chitinase in chickpea roots with resistance to *Fusarium oxysporum* f. sp. *ciceri*. *Plant Pathology*, 47, 693-699.

12. Guo, D., Chen, F., Inoue, K., Blount, J. W., and Dixon, R. A. (2001). Down regulation of caffeic acid 3-O-methyltransferase and caffeoyl CoA 3- O-methyltransferase in transgenic alfalfa: impacts on lignin structure and implications for the biosynthesis of G and S lignin. *Plant Cell* 13, 73–88.

13. Iannone, M.F., Groppa, M.D., Benavides, M.P. (2015). Cadmium induces different biochemical responses in wild type and catalase-deficient tobacco plants. *Environ. Exp. Bot*. 109, 201–211.

14. Jayaraman, K. S., Ramanuja, M. N., Vijayaraghavan, P. K., and Vaidyanathan, C. S. (1987). Oxidative enzyme in pearl millet. *Food Chem*, 24(3), 203-206.

15. Kamble, P.N., Giri, S.P., Mane, R.S., and Tiwana, A. (2015). Estimation of chlorophyll content in young and adult leaves of some selected plants. *Universal journal of environmental research and technology*, 5(6), 306-310.

16. Kumar D, Yusuf M, Singh P, Sardar M, Sarin N. (2014). Histochemical Detection of Superoxide and H_2O_2 Accumulation in *Brassica juncea* Seedlings. Bio-Protocol. 4.

17. Lee, C. P., Liu, P. T., Kung, H. N., Su, M. T., Chua, H. H., Chang, Y. H., Chang, C. W., Tsai, C. H., Liu, F. T. and Chen, M. R. (2012). The ESCRT machinery is recruited by the viral BFRF1 protein to the nucleus-associated membrane for the maturation of *Epstein-Barr Virus*. *PLoS Pathog* 8(9): e1002904.

18. Masoodi, K.Z., Lone, S.M. and Rasool, R.S. (2020). *Advanced Methods in Molecular Biology and Biotechnology: A Practical Lab Manual*. Academic Press.

19. Mittler, R., and Zilinskas, B. A. (1992). Molecular cloning and characterization of a gene encoding pea cytosolic ascorbate peroxidase. *Journal of Biological Chemistry*, 267(30), 21802-21807.

20. Prassan, C., Shaloo, V., Sudipta, D., and Hillol, C. (2019). National training on Biocontrol strategies for invasive and emerging plant pathogens. ICAR-NBAIM, Kushmaur Mau.

21. Rio, D. C., Ares, M., Hannon, G. J., and Nilsen, T. W. (2010). Purification of RNA using TRIzol (TRI reagent). *Cold Spring Harbor Protocols*, 2010 (6), pdb-prot5439.

22. Rotruck, J.T.; Pope, A.L; Ganther, H.E; Swanson, A.B; Hafeman, D.G and Hoekstra, W.G. (1973). Selenium: Biochemical role as a component of glutathione peroxidase purification and assay, Science, 179, 588-598.

23. Sadasivam, S., and Manickam, A. (1996). *Biochemical methods* (2nd ed.) New Delhi: New Age International.

24. Sairam, R.K., and Srivastava, G.C. (2002). Changes in antioxidant activity in sub- cellular fractions of tolerant and susceptible wheat genotypes in response to long term salt stress, *Plant Science*,162, 897-904.

25. Sidhu, J. S., Ravi, J. S., and Minocha, J. L. (1984). Peroxidase isozyme patterns in primary trisomics of pearl millet. *Theoretical and Applied Genetics*, 68(1), 179-182.

26. Singh, S., Singh, U.B., Malviya, D., Paul, S., Sahu, P.K., Trivedi, M., Paul, D. and Saxena, A.K. (2020). Seed biopriming with microbial inoculant triggers local and systemic defense responses against *Rhizoctonia solani* causing banded leaf and sheath blight in maize (*Zea mays* L.). *International Journal of Environmental Research and Public Health*, 17(4), 1396.

27. Singh, U. B., Singh, S., Malviya, D., Chaurasia, R., Sahu, P. K., Sharma, S. K., and Saxena, A. K. (2020). *Drechslerella dactyloides* and *Dactylaria brochopaga* mediated structural defense in tomato plants pre-challenged with *Meloidogyne incognita*. *Biological Control*, 143, 104202.

28. Thordal-Christensen, H., Zhang, Z., Wei, Y., Collinge, D.B. (1997). Subcellular localization of H_2O_2 in plants, H_2O_2 accumulation in papillae and hypersensitive response during the barley-powdery mildew interaction. *Plant J.*, 11, 1187–1194.

29. Woodbury W, Spencer AK, Stahmann MA. (1971). An improved procedure using ferricyanide for detecting catalase isozymes. *Anal Biochem*. 44: 301-305.

30. Zhang, N., Yu, D., and Zhu, X. (2018). RNA Isolation from Plant Tissues: A Hands on Laboratory Experimental Experience for Undergraduates. *Biochemistry and Molecular Biology Education*, 46(3), 253-261.

Index